Charles Sutherland

Horses, asses, zebras, mules and mule breeding

Charles Sutherland

Horses, asses, zebras, mules and mule breeding

ISBN/EAN: 9783337142988

Printed in Europe, USA, Canada, Australia, Japan

Cover: Foto ©berggeist007 / pixelio.de

More available books at **www.hansebooks.com**

HORSES, ASSES, ZEBRAS, MULES,

AND

MULE BREEDING.

BY

W. B. TEGETMEIER, M.B.O.U., F.Z.S.,

AND

C. L. SUTHERLAND, F.Z.S.,

LATE OF THE WAR OFFICE; ATTACHED TO THE BRITISH COMMISSION PHILADELPHIA EXHIBITION, 1876; INTERNATIONAL JUROR PARIS EXHIBITION, 1878; ASSISTANT COMMISSIONER, ROYAL COMMISSION ON AGRICULTURE, 1879.

LONDON:

HORACE COX,

"FIELD" OFFICE, WINDSOR HOUSE, BREAM'S BUILDINGS, E.C.

1895.

LONDON:
PRINTED BY HORACE COX, WINDSOR HOUSE,
BREAM'S BUILDINGS, E.C.

PREFACE.

Upwards of four thousand works on horses and their utilization have been published, and of this number about one half have been printed in Great Britain. It may therefore appear an act of presumption on the part of any writer to augment the already lengthy list, but recently new animals, such as Prejevalski's horse and Grevy's zebra, have been discovered; species hitherto untamed have been pressed into the service of man, and new hybrids have been reared which hold out the promise of great utility.

Much knowledge has been gained by recent travellers respecting the history and habits of species hitherto imperfectly known, and, above all, a vast amount of information has been accumulated, proving the advantages that are found to arise from utilizing the mule in almost all civilized countries excepting England, in which country no book on this useful hybrid has ever been published. To supply this deficiency; to demonstrate the great value and economy of the mule as a beast of draught and burden, that could be as advantageously employed in this country in agricultural and general draught purposes as it is by other nations, and by ourselves in all military operations abroad, is in part the object with which this work is published.

CONTENTS.

PART I.

HORSES, ASSES, AND ZEBRAS.

CHAPTER I.

The Horse ... ... *page* 1

CHAPTER II.

Prejevalsky's Horse 7

CHAPTER III.

African Wild Ass 11

CHAPTER IV.

Wild Ass of Somaliland 19

CHAPTER V.

Asiatic Wild Ass 21

CHAPTER VI.

Mountain Zebra 37

CHAPTER VII.

Grevy's Zebra 43

CHAPTER VIII.

Burchell's Zebra 51

CHAPTER IX.

The Quagga 61

CHAPTER X.

Hybrid Equidæ 65

PART II.

MULES AND MULE BREEDING.

CHAPTER XI.
The Utilisation of Mules ... ... page 71

CHAPTER XII.
Non-fertility and Lactation in Mules 79

CHAPTER XIII.
The Poitou Mule 85

CHAPTER XIV.
Poitou Ass as a Sire of Mules 95

CHAPTER XV.
American Mule 107

CHAPTER XVI.
Mules for Military Service 127

CHAPTER XVII.
Practical Remarks on the Use of Mules 138

APPENDIX.
MEMORANDUM ON MULE BREEDING IN INDIA.

LIST OF ENGRAVINGS.

		PAGE
Italian Razza Jack ...	*Frontispiece*	
Prejevalsky's Horse	... facing	7
African Wild Ass	„	11
Somali Wild Ass	„	19
Onager	... „	23
Captured Onagers at Morvi	„	27
Kiang	„	31
Mountain Zebra	„	37
Zebra Broken to Saddle (in text)		41
Grevy's Zebra	... „	43
Skin of Grevy's Zebra (in text) ...		44
Burchell's Zebra	... „	51
Skin of Burchell's Zebra (in text)	...	53
Utilisation of Burchell's Zebra	... „	55
Burchell's Zebra in Cape Cart	... „	57
Burchell's and Mountain Zebra Contrasted ...	... „	59
Quagga	... „	61
Burchell's Zebra and Hybrid (in text)		68
Supposititious Mule	„	81
Brown Poitou Mule	... „	85
Poitou Mule, Brunette	... „	89
English Mule	... „	93

	PAGE
Poitou Jack, Typical ... ... facing	95
Poitou Jack, Smooth-coated ... ... „	99
Poitou Jack, Yearling ... ... „	101
Poitou Jenny ... ... „	103
Poitou Mare and Mule Foal ... ... „	105
Gun Mule, Indian Mountain Battery ... „	129
Packing Baggage Mules (in text)	132–133

PART I.

HORSES, ASSES, AND ZEBRAS.

CHAPTER I.

THE HORSE.

(*Equus caballus. Linn.*)

In the views of modern zoologists all the species of wild horses, asses, and zebras at present existing constitute but one genus, distinguished by the name Equus, the separation of the asses and zebras into distinct genera under the names of Asinus and Hippotigris, as has been proposed by some zoologists, not being generally accepted.

The number of the existing species of the genus Equus cannot be accurately defined, but may be taken as not exceeding twelve in number.

1. Equus caballus (the Horse).
2. Equus przewalskii (Prejevalsky's horse).
3. Equus asinus (African wild ass).
4. Equus asinus somalicus (Somali wild ass).
5. Equus onager (the Hemione).
6. Equus kiang (the Kiang).
7. Equus hemippus (the Hemippe).
8. Equus zebra (the Mountain zebra).
9. Equus burchellii (Burchell's zebra).
10. Equus chapmanii (Chapman's zebra).
11. Equus grevyi (Grevy's zebra).
12. Equus quagga (the Quagga).

Several of these, it is probable, are mere local varieties,

or what naturalists term sub-species. This is possibly the case with Prejevalsky's horse, and may be so with the Somali ass. The kiang and the hemippe are now regarded by naturalists as local varieties of the onager; Chapman's zebra appears to be but a variation of the older known Burchell's zebra; and the quagga is now generally believed to have been exterminated.

The horse is distinguished from the other Equidæ by the tail being covered with long hair from its base to its end; it possesses also a small bare callus on the inner side of the hind leg below the hock, in addition to the one on the inner side of the foreleg, which is present in all the other species. Further distinctions are the longer mane and forelock, and shorter ears, whilst, in proportion to its size, its limbs are longer, its hoofs broader, and its head smaller than in the species known as wild asses and zebras.

An important distinction between the horse and the other species of the genus, namely, the asses and zebras, appears to have been overlooked or mis-stated by preceding writers—namely, the difference in the period of gestation; this in the horse is eleven months, whilst in the asses and zebras it exceeds twelve months, as evidenced in the succeeding chapters on those animals.

It is remarkable that this difference should have been so generally ignored. Thus, it is not mentioned by Capt. Hayes in his "Points of the Horse," although he devotes no less than five pages to the enumeration of the "Differences between the Ass and Horse." Again, Mr. Blanford, in his "Fauna of British India: Mammalia," writing of the Asiatic wild ass, says: "The period of gestation is probably the same as in the horse and ass, about eleven months," and Sir William Flower, in his "Mammals, Living and Extinct," when describing the general characters of the

Equidæ, states definitely "the period of gestation is eleven months."

The distribution of the horse on the earth's surface at the present time is largely owing to the agency of man. In Europe wild horses were exceedingly abundant long before the historic period. Their remains are found associated with those of man and domesticated animals belonging to what is called the Neolithic or Stone period. Representations of horses have been found carved on bones and antlers in caves in the south of France, the horse resembling that which is at present feral in the Steppes of Russia and Tartary. Cæsar records the ancient Britons as using war chariots, and the horse is represented in the monumental records of Egypt nearly 2000 years before the Christian era. Horses have now been conveyed to every part of the civilised world. It is probable, though not quite certain, that the horse did not exist in the historic period in America until after its discovery by Columbus, although it is remarkable that fossil remains of true horses are found in almost every part of America. They then appear to have been exterminated and have since been re-introduced by man, and have now become feral in large numbers. The horse was undoubtedly introduced by man into Australia, no hoofed animals existing in that vast continent at the time of its discovery.

Whether there are any truly wild horses at the present time—that is to say, animals whose ancestors have never been domesticated—is doubtful. Sir William Flower says that the nearest approach to the truly wild horse existing at the present time are the so-called tarpans, which occur in the Steppe country north of the Sea of Azoff. They are small in size, dun colour, with short mane, and rounded obtuse nose. There is no evidence to prove whether they are really wild—that is, descendants of animals which have never been domesticated; or feral—that is, descended from animals which have escaped from captivity, like the horses that roam over the plains of South America and Australia, and the wild boars that now inhabit the forests of New Zealand.

Enthusiastic as sportsmen and hunting men may be over the form and endowments of the horse, it is hardly too much to say that naturalists enjoy the contemplation of this glorious creature with no less pleasure, tracing with great interest the modifications that have taken place from the forms of the old extinct horse-like animals as shown in their fossil remains; modifications which have adapted the modern horse to the present condition of things on the earth's surface. The extinct horse-like animals of the older world had large feet with three and even four toes, with short legs adapted for walking on marshy or yielding ground, like the tapirs and rhinoceroses of modern times. Leaving out of consideration these extinct animals, and speaking of the modern horses only, we find that the specimens of the genus Equus are inhabitants of the plains, for which their whole organisation is specially adapted.

It is interesting to compare the statements of two of the most eminent zoologists regarding these equine animals. The late Sir Richard Owen, in his "Anatomy of Vertebrates," writes most graphically on the fitness of the organisation of the horse for the needs of man, and he speaks of the coincidence of the modification of the old fossil forms into the present animals with the earliest evidence of the human race. He fervently descants on the fact that, of all the servants of man, none have proved of more value to him. The horse, he says, since its subjugation, has acquired nobler proportions, higher faculties, more strength, more speed, and more amenability to guidance.

"No one (writes Sir Richard Owen) can enter the 'saddling ground' at Epsom, before the start for the Derby, without feeling that the glossy-coated, proudly-stepping creatures led out before him are the most perfect and beautiful of quadrupeds. As such, I believe the Horse to have been predestined and prepared for Man. It may be weakness, but, if so, it is a glorious one, to discern, however dimly, across our finite prison wall, evidence of the 'Divinity that shapes our ends,' abuse the means as we may."

Sir William Flower, the successor to Sir Richard Owen, in describing the horse from a somewhat different standpoint, speaks of the adaptation of its organisation to its life on the open plains, where it is found. He calls attention to the length and mobility of the neck, the position of the eye and ear, the great development of the organ of smell (which gives the wild horses, asses, and zebras the means of becoming aware of the approach of their enemies), while the length of their limbs, the angles which the different segments form with each other, and the combination of firmness, stability, and lightness resulting from the reduction of all the toes to a single one, upon which the whole weight of the body and all the muscular power are concentrated, give them speed and endurance surpassing that of almost any other animal.

"If we were not so habituated (writes Sir William Flower) to the sight of the horse as hardly ever to consider its structure, we should greatly marvel at being told of a mammal so strangely constructed that it had but a single toe on each extremity, on the end of the nail of which it walked or galloped. Such a conformation is without a parallel in the vertebrate series, and is one of the most remarkable instances of specialisation, or deviation from the usual type, in accordance with particular conditions of life."

The consideration of the varieties of the horse, which have resulted from its long domestication, does not come within the scope of the present work, except as far as the different breeds influence the character of hybrids between the horse and the other species, a subject that will be fully considered in the concluding chapters on Mules and Mule Breeding.

Prejevalsky's Horse (*Equus przewalskii*).

CHAPTER II.

PREJEVALSKY'S HORSE.

(*Equus przewalskii. Poliakof.*)

MUCH interest has been excited amongst naturalists respecting the existence of a supposed additional species of horse, which was first brought to notice by, and subsequently named after, the distinguished Russian traveller Prejevalsky.* His single specimen, which he presented to the Zoological Museum of the Imperial Academy, St. Petersburg, was not really captured by him, but was given to him by the chief magistrate of the district of Zaisan, it being at that time the only one that had been obtained by the wild camel hunters in the deserts of central Asia. A drawing of this animal was published by Prejevalsky, and is accurately reproduced in the engraving by Mr. Frohawk. The specimen was described at considerable length by the Russian naturalist Poliakof in the "Proceedings of the Russian Geographical Society" for January, 1881. This description was translated by Mr. E. Delmar Morgan, and published in the "Annals and Magazine of Natural History" for 1881. Poliakof distinguishes the animal from the tarpans or

* In the above account I have employed the western mode of spelling Prejevalsky's name, retaining the Russian form only when used as the specific appellation, which, in accordance with the rules governing scientific nomenclature, I cannot alter.

so-called wild horses of Tartary, which appear to be really domestic horses that have recovered their liberty, and maintains that it is a perfectly distinct species. In his description he says that the specimen is about three years old, its size is equal to that of the wild asses, but that its head is better shaped near the end of the muzzle, and has shorter ears than those of the wild ass. In shape it takes after the horse, the legs being relatively thick for the size of the body, the hoofs round and broader, and the tail better furnished with hair than the wild ass. The colour is dun, with a yellow tinge on the back, becoming lighter towards the flank and almost white under the belly. The hair is long and wavy, brick-red on the head, cheeks, and lower jaw. The extremity of the nose is almost covered with white hairs, in strong contrast to the red of the other parts of the head. It has no forelock, but the mane is short, upright, and "hogged," extending from between the ears to the withers, and of a dark brown colour. There is no dorsal stripe along the back, as in the Asiatic asses. The upper half of the tail is the same colour as the back, but it is longer and thicker at the root than that of any kind of ass. The extremity of the tail is covered with dark brown, or nearly black, hair. The fore legs are brown near the hoofs and on the knees, a peculiarity, he says, which is never known to occur with wild asses, and dark hairs occur on the lower part of the hind legs. The skull and the hoofs more closely resemble those of the horse than any animal of the asinine group. Such is Poliakof's description of the animal; commenting on which Sir William Flower writes as follows:

"It is described as being so intermediate in character between the equine and the asinine group of Equidæ, that it completely breaks down the generic distinction which some zoologists have thought fit to establish between them. It has callosities on all four limbs, as in the horse, but only the lower half of the tail is covered with long hairs, as in the ass. The general colour is dun, with a yellowish tinge on the back, becoming lighter towards the flanks, and almost white under the belly, and there is no dark dorsal stripe. The mane is dark brown, short, and erect, and there is no forelock. The hair is long and wavy on the head, cheeks, and jaws. The skull and the hoofs are described as being more like those of the horse than the ass. Until more specimens are obtained, it is difficult to form a definite opinion as to the validity of this species, or to resist the suspicion that it may not be an accidental hybrid between the kiang and the horse."

Additional specimens of this interesting animal have recently been obtained. The Brothers Grijimailo met with this wild horse in the desert of Dzungaria. The account of their expedition, which was published in the "Proceedings of the Russian Geographical Society," has been translated, with notes by Mr. E. Delmar Morgan, and published in the "Proceedings of the Royal Geographical Society" for April, 1891. In their account they state as follows:

"Springs enable the numerous animals inhabiting Dzungaria to exist; of these the most interesting is Prejevalsky's horse (*E. przewalskii*). The only known specimen of this animal, in the Zoological Museum of the Imperial Academy, was obtained by Prejevalsky from the chief magistrate of the district of Zaisan, who had received it from the Kirghiz. Prejevalsky himself, though he crossed the desert of Dzungaria in three several directions, never came across any of these wild horses, and if he wrote otherwise he was mistaking kulans (*E. onager*) he had seen in the distance for wild horses, a mistake the most experienced hunters are liable to make, for at that distance it is almost impossible to distinguish between them. It is only by their manner of holding themselves that these animals may be recognised. The stallion of the wild horse never leads the herd, but is always behind, taking care of the young, which he protects better than do the mares. But however this may be, we were the first Europeans who, for twenty days, made a study of these interesting animals, adding the skins of three handsome stallions and one mare to our collection—an acquisition we may well be proud of, though made at the cost of many hardships and privations. Besides Equus przewalskii, Dzungaria has the tiger, two antelopes (*A. saiga* and *A. gutturosa*), two wild asses (*E. hemionus* and *E. onager*), and, among small animals, a hare and a few rodents not yet determined."

They seem to throw no suspicion on Prejevalsky's horse being a distinct species, and do not even allude to the possibility of its being a hybrid between the ass and the horse. They obviously paid great attention to their zoological collection, having obtained a large number of specimens, comprising twenty-nine large mammalia, thirty-nine medium, and forty-two small, and they regarded the four specimens of Equus przewalskii as amongst their most interesting acquisitions. Of these they secured three skulls and one perfect skeleton. Sir W. Flower thinks that it is difficult to form a definite opinion as to the validity of this species, or resist the suspicion that it may be a mule. The latter supposition is unlikely, as, if it were true, so many specimens could hardly have been obtained; moreover, hybrids between two species are rarely produced except through the agency of man. The capture of a female E. przewalskii in foal would settle this disputed question, equine mules being invariably barren.

AFRICAN WILD ASS (*Equus asinus*).

(From specimen in Zoological Gardens, 1893.)

CHAPTER III.

THE AFRICAN WILD ASS.

(*Equus asinus. Linn.*)

The two species of the genus Equus, namely Equus caballus and Equus przewalskii, are both regarded as horses, being distinguished, amongst other characters, by the presence of callosities, also known as ergots, chesnuts, or castors, on both the hind and fore legs; these are absent from the hind legs of the other species, whilst some of the hybrids (mules) have them, and others have not—by their broad hoofs, and by the long hair not being confined to the extremity of the tail.

The remaining equine animals may conveniently, though not with any great accuracy, be divided into two groups, those which are plainly coloured, the true asses, and those which are striped, and are known popularly as zebras.

The distinction, though obvious to the eye, has no great zoological value. Several of the varieties of the horse, such as the pure bred Norwegian ponies, habitually have the spinal and leg stripes, and numerous other breeds that possess them are described by Darwin in his work on "Variation," and one African ass, that from Somaliland, is characterised by its transverse leg stripes. The asses are, however, characterised by their geographical distribution, those from Africa being markedly distinct from the Asiatic species.

The African wild ass is now regarded by all zoologists as the origin of our domesticated animal. As this species was originally termed Equus asinus by Linnæus, the name should be retained in place of Equus tœniopus, which was subsequently given it by Heuglin, even although the latter has been extensively used in scientific works. The appellation tœniopus, stripe footed, expresses the fact that many of the species possess dark markings on the lower part of the limbs.

The African ass is found wild in Abyssinia, Nubia, and other parts of North-east Africa lying between the Nile and the Red Sea. Its colour and markings approach closely to those of the ordinary domestic ass; it possesses a distinct shoulder stripe running from the withers down to the commencement of the fore leg, similar to that seen almost invariably in the donkey. The ears of the African are longer than those of the Asiatic asses. The activity and speed of the animal must not be judged of by those of the domesticated ass, which has suffered, in this country at least, from continued neglect and scanty fare for centuries. Sir Samuel Baker, speaking of the wild ass, says:

"Those who have seen donkeys only in their civilised state can have no conception of the beauty of the wild or original animal. It is the perfection of activity and courage. It has a high bred tone in its deportment, a high-actioned step when it trots freely over the rocks and sand, with the speed of a horse when it gallops over the boundless desert. The specimens now in the Zoological Gardens will enable any one to perceive the character of the animal as it was before being altered by generations of captivity."

The bray of the African is identical with that of our common ass, and Darwin, in his "Variation," notes the marked aversion to walking across a brook, which characterises our domestic donkey, as indicating its being derived from a desert-haunting animal, as also, he says, does its pleasure in rolling in the dust.

Of the African wild ass there are now (1894) three specimens in the Zoological Gardens. A female purchased by the Society in 1881, has repeatedly bred, once with the Asiatic ass (*E. hemionus*), and four times with a male of her own species.

The male African ass now in the Regent's Park is stated to be not a native of Africa, but is said to have come from the island of Diego Garcia, in the Indian Ocean, where African asses were taken by sailing vessels, and have become wild, retaining all their characteristics, although somewhat reduced in size. In reference to this statement, Admiral Kennedy, of H.M.S. "Boadicea," writing from Madagascar, June 25, 1893, informed me that he had lately met with a gentleman who had lived for eighteen years on the island of Diego Garcia, during which time he had never met with a donkey (at least a wild one), and he is certain that such an animal never existed there.

As is the case with the horse, the ass has been so long a period under domestication that great variations exist in its size and general character. Some asses in India are said to be not larger than Newfoundland dogs. In the south of Europe they are reared with care and attain a large size, and in Poitou a very large breed, of great strength and stoutness of limb, is reared for the purpose of breeding draught mules that rival our large draught cart horses in size and strength.

The period of gestation in the ass is not, as generally stated, even in scientific works such as Blanford's "Fauna of British India: Mammalia," identical with that of the horse. The stud book of Mr. C. L. Sutherland, who is so greatly interested in ass and mule breeding, testifies to the fact that it is a month longer, and in some instances even more. Thus, a Spanish jenny that visited a Poitou jack on July 7, 1876, foaled a jack on July 23, 1877.* Another distinction between the ass and the horse is that twin births are not uncommon in the case of the former, but are much rarer in the latter.

A singular variety of the African ass was foaled in the Gardens of the Zoological Society in 1892. The female parent was purchased by the society in 1881, and produced her first foal in June, 1883; the other parent being not an African, but the variety of the Asiatic ass known as the Hemippe (*E. hemionus*). This offspring was of a reddish colour, similar to that of the male parent, the female being of the usual grey. This half-bred was exhibited at the Agricultural Show at Windsor, in 1889. It was a vicious, untamed animal, and was sold to Mr. Guy, and is now in the Gardens of the Zoological Society at Dublin. In the report of that Society

* As this difference in the gestation of the two species is so generally ignored, I have thought it desirable to adduce the following definite instances of the period of gestation in the ass (the result in six cases of a single service) from the stud book of Mr. Sutherland:

Dam.	Sire.	Date.	Foal Born.	Period of Gestation.
Donna	to Ranulfe......	May 23, 1877	June 12, 1878 (Jenny) ...	385 days
Dolores	to Ranulfe......	April 19 and 20, 1877 ...	April 21, 1878 (Jack) ...	366 days
Adéle	to Vitré	June 24, 1879	June 16, 1880 (Jack)......	358 days
Donna	to Vitré	June 23, 1879	June 26, 1880 (Jenny) ...	369 days
Donna II.	to Vitré	Sept. 30 & Oct. 27, 1879...	Oct. 13, 1880 (Jenny) ...	379 days (?)
Nellie	to Don Juan...	April 27, 1885	May 16, 1886 (Jack)	385 days
Dinah	to Cetywayo...	June 15, 1887	June 14, 1888 (Jack)......	365 days
Nellie	to Malta Jack.	June 18, 1886	July 3, 1887 (Jenny)......	380 days

The above were all large jennies of foreign breeds, such as the Maltese, Spanish, Poitou, and their crosses, which belonged to Mr. Sutherland. They were put to large foreign jacks with the above results.

for 1892 this hybrid is described as a fine animal, which resembles its male parent rather than the mother. Since then the female African ass has produced other foals, the male parent being in all cases an African of the usual grey colour and dark shoulder stripe common to the species. These foals were born respectively in 1889, 1891, and the last on October 13th, 1892. The latter offers a striking example of variation from the usual markings and colour of the species to which it belongs. It is of a reddish fawn colour, somewhat shaggy in coat, and is remarkably distinguished by a large star on the forehead and a white blaze down the face, such as is rarely, if ever, seen in any species of ass, wild or domesticated. There is the slightest possible indication of stripes on the legs and of the shoulder stripe, and the ears are shorter than those of the parents. It has passed into the possession of Mr. A. J. Scott, of Rotherfield Park, Alton, Hants, and is remarkably tame and sociable.

Now, the question that presents itself is whether this is an accidental variation, such as occurs from time to time in almost all animals, especially those in confinement or domestication, or whether it is an instance of the influence of a previous impregnation, and that the animal has reverted to the characters of the Hemippe, which was the parent of the first foal produced by the female.

The influence of the first sire on all subsequent offspring is a subject of very considerable importance that has not received the scientific investigation that it merits. It is generally accepted by breeders of dogs, and in the case of valuable animals the effect of a *mésalliance* is carefully guarded against. It is one that is recognised by physiologists as affecting the human species, and the example of the striped foals that were always bred by a mare whose first foal was a hybrid of zebra parentage is well known.

Whether this young ass bred in the Gardens is merely an accidental variation, or whether it owes its peculiarities to the influence of the Hemippe, is a point which I will not endeavour to decide. The white blaze on the face is most peculiar, and I am informed by Mr. C. L. Sutherland, who is well known in connection with the breeding of equine animals and their hybrids, that he has never seen this blaze, so common on the horse, on any of the many thousand asses that have come under his notice in Europe and America. The facts of the case are, therefore, worth putting on record.

Captain Hayes, in his recent work on "The Points of the Horse," says:—

"The ass hardly ever has any irregular markings on its coat, such as a 'star,' 'blaze,' 'reach,' or 'stockings,' all of which are very frequent amongst horses. A small star, on one or two occasions, is the only mark of the kind I have ever seen in the ass. At the same time, I must state that I have not had much experience among these animals.

"I believe I am correct in saying that the colour of the ass is never of a bright bay, chestnut, red or blue roan, or nutmeg grey. I have seen mules of an iron-grey colour, but have not observed it in the ass. This conservatism in colour and freedom from irregular markings, shown by the ass, is very remarkable, considering how greatly the coat of the horse varies in this respect."

Captain Hayes also calls attention to the different extent of the patches of thickened skin, which he terms the shell, that cover the croup and the pelvis in the horse, whereas in the ass it extends all over the ribs, which are consequently not as sensitive to the effects of blows as are those of the horse. This thickening is due to an extremely dense layer of connective tissue, which is so close and hard that when the skin has been tanned and dried it looks like horn, and is utilised for the manufacture of the long boots worn by foreign cavalry officers.

Mr. C. L. Sutherland furnishes me with the following list of five cardinal points in which the ass differs from the horse:

"(1) In the period of gestation, which in the ass, as before tated, is at least twelve months, whereas in the horse it is eleven.

"(2) The absence of chestnuts on the hind legs of the ass.

"(3) The number of loin vertebræ in the ass is five, in the horse six. In the mule it is sometimes five and sometimes six.

"(4) The ass in comparison with the horse more frequently produces twins; in many cases, however, these are the result of superfœtation, as is evidenced by the difference in size of the produce. In my experience, writes Mr. Sutherland, an ass in foal with twins always aborts.

"(5) The entire absence in the ass of the white stockings or fetlocks so common in the horse, and also of the star or blaze on the forehead. In these particulars the mule follows the ass, which is very prepotent over the horse in those cases where the ass is the male parent. The mule may be said to be three-fourths of an ass rather than intermediate between its parents, whereas the mute, which is also called a hinny or jennet (the 'bardot' of the French), in which the horse is the male parent, favours the horse rather than the ass.

"Piebald or skewbald asses, though sometimes occurring, are not common, and can only be produced from parents of which one at least is either piebald or skewbald. A white jack and a brown jenny, or the converse, will not produce broken coloured offspring, unless this character has previously appeared in the ancestors of one or other of the parents."

A further distinction between the two species exists in

the striking difference in the longevity of the ass as compared with that of the horse; the latter rarely attains the age of twenty-five years, whilst asses of thirty years are not infrequent, and instances of much greater longevity are on record; thus, in the *Graphic* of July 1, 1893, is given a portrait of a donkey now living on which, it is stated, the present Earl of Feversham used to ride fifty-five years ago.

SOMALI WILD ASS (*Equus asinus somalicus*).

(From specimen in Zoological Gardens, 1884.)

CHAPTER IV.

THE WILD ASS OF SOMALILAND.

(*Equus asinus somalicus, Sclater.*)

THIS animal was brought under the notice of the Fellows of the Zoological Society in November, 1884, when the secretary, Mr. P. L. Sclater, described and exhibited a skin, and called attention to a fine specimen then living in the Gardens, having been deposited by Mr. Hagenbeck. Mr. Sclater at the same time called attention to another African wild ass (*E. asinus*), from the Nubian desert, which was purchased in May, 1881, and compared the two, demonstrating that they belonged to distinct species or sub-species.

As will be seen by the engraving from Mr. Smit's drawing, which appeared in the *Proc. Zoolog. Soc.*, 1884, the Somali ass differs from the ordinary African wild ass in its more greyish colour, in the entire absence of the cross-stripe over the shoulders, in the very slight indication of the dorsal line, and more especially in the numerous black markings on both front and hind legs. It has, likewise, smaller ears, and a longer mane.

These cannot be regarded as individual variations, for they were present in the skin from Somaliland, which was exhibited at the same time by Mr. Sclater. Moreover, Mr. E. Lort Phillips, who visited the Berberah district in March, 1884, ascertained that the wild asses which he there met with were all precisely of the same description.

Mr. Lort Phillips wrote as follows:

"On March 22, 1884, when about twenty miles to the west of Berberah, we fell in with a small herd of wild asses. After a long and tedious stalk I succeeded in bagging one, which turned out to be of quite a new species to me, having no mark whatever on the body, which was of a beautiful French grey colour. On its legs, however, it had black stripes running diagonally. I have unfortunately lost the book in which I put its measurements, but it was a superb creature, and stood quite 14 hands at the shoulder; our Berberah horses looked quite small in comparison."

Whether this Somali ass should be regarded as a distinct species from the ordinary African wild ass or merely a local variation is uncertain, and depends on the view taken of specific distinctions by each individual. There appears to me to be little doubt they would breed together and produce fertile offspring. It is not without interest to remark that as we go further south from Abyssinia towards the Cape the asses approximate more closely to the striped equines, the zebras and quaggas of South Africa.

CHAPTER V.

THE ASIATIC WILD ASS.

(*Equus hemionus, Pallas*).

At the present time naturalists incline to the opinion that there is but one distinct species of Asiatic wild ass, to which they assign the name of Equus hemionus, first bestowed on it by Pallas. In the list of animals that have been exhibited in the gardens of the Zoological Society, three species of the Asiatic wild ass are enumerated, named respectively—(1) the Asiatic wild ass (*E. onager*), the ghor-khur of Western India and Baluchistan; (2) the Hemippe (*E. hemippus*) from Persia and Syria; and (3) the Kiang (*E. hemionus*) from Tibet. The first two differ but very little from one another, but the kiang or dzeggetai is stated by Mr. Blanford to be darker and redder than the ghor-khur, and to have a narrower dorsal stripe, although he agrees with Sykes, Blyth, and Flower in regarding these three wild asses as constituting but varieties of one species.

The Asiatic wild ass inhabits the vast open plains that exist in various parts of Asia, from Syria through Persia, Afghanistan, the Punjab, and Tibet, right away to the frontiers of China. It is usually found in herds varying in number from four or five to thirty or forty individuals. In the spring months the mares and foals sometimes collect in vast numbers, and Dr. Aitchison, in his report on the

Afghan frontier expedition of 1884, stated that he saw a herd in north-western Afghanistan that he regarded as consisting of some thousand animals.

In the description of this species by Blanford in his "Fauna of British India," we are informed that the ears are large, that the tail is covered with short hair near the base, which grows gradually longer towards the end, that the mane is erect, and that there is a naked callosity on the inside of each fore-arm, but none on the hind legs. The general colour of the Asiatic wild ass is a sort of rufous grey, which varies to fawn colour, or even pale chestnut. The under parts of the body are white. A dark brown stripe, which varies in breadth, sometimes being margined with white, runs down from the nape to the tail, and occasionally there is a dark cross stripe on the shoulder, and faint rufous bars are said to occur at times on the limbs. The end of the tail is blackish. In height the Asiatic wild ass varies from 3ft. 8in. to 4ft. Its food consists of various grasses and the herbage of other plants. Its voice is described by Mr. Blanford as being a shrieking bray. These wild asses are remarkable for their speed and endurance. Mr. Blanford informs us that in the country west of the Indus, the mares are said to drop their foals in June, July, and August; the period of gestation he regards as eleven months, but it is more probably identical with that of the African species, and exceeds twelve months. Two local varieties of the Asiatic ass, the Onager and the Kiang, though not regarded by zoologists as specifically distinct, call for distinct notice.

The Onager of India (*Equus hemionus*, var. *E. onager*).

(From photograph taken in India.)

THE ONAGER.

GHOR-KHUR, Hindi; GHOUR or KHERDECHT, Persian; KOULAN, Kirghiz; GHUR, GHURÁN, Baluch.

THE geographical race or variety of the Asiatic ass usually spoken of as the Onager is well illustrated in the engraving from a specimen sent to the Zoological Gardens in 1873 by Captain Henry Lowther Nutt, who, writing to the secretary of the Zoological Society, stated:

"I ran it down on the Runn of Kutch. I was riding hard after it for three hours and five minutes, and the estimated length of the chase was forty miles. I rode two horses, as I discovered from the 'puggies,' or watchers near the Runn, that, if the animals were disturbed from where they were, they would probably make for another place some twelve or thirteen miles distant. I was, therefore, able to post a fresh party of horsemen, and a fresh horse for myself, at the place further on; and true enough the herd did make for the spot indicated, so that the running was taken up and continued with fresh horses, and in this way the capture was effected, but even then not until both my horses, which were in good order at the time, had been ridden to a standstill. This will give you an idea of the speed and endurance of these animals."

A full account of the exciting chase of this specimen was published by Captain Nutt in the *Oriental Sporting Magazine*, under the title of "Donkey Hunting on the Runn of Kutch." The engraving of the animal with its syce was copied from a photograph forwarded from India by Mr. Fraser S. Hore, in whose possession the animal remained some time previous to its embarkation. Mr. Hore, in a letter to Mr. C. L. Sutherland, writes as follows:

"I send you two photographs and an account of the animal in

its wild state. The specimen in question was ridden down on the Runn of Kutch in the month of March, 1873. She is the only instance known (bar one, when the beast was on the point of dropping a foal) of a wild ass having been run down before. The party that captured her was headed by a friend of mine, Capt. H. Lowther Nutt, Acting Second Political Assistant, Kattyawar. The photographs I had taken myself, and the donkey at present is in my possession, waiting to be conveyed to the Zoological Society by the first Suez Canal steamer that leaves Bombay, the society having provided the funds for its passage home. Its age at present time (October, 1873) is about one year; its colour is a mixture of white and fawn; the under parts of the body, the neck and chest, nose and nasal region, back part of face, rump, channel and inside of the legs are white; the mane is short, stumpy, and dark brown. A dark dun streak of longish hair runs down the back, broadening towards the rump, and continuing down the tail to the end. The other parts of the body and head are of a fawn colour, the entire coat being smooth and glossy; the tail has a small tuft of long dark-brown hair at the end. The legs are beautifully clean and flat, the back sinews standing well out; and there is a black, shiny, horny ergot high up inside each fore leg; the feet are beautifully formed, hard, and very small; pasterns very long on fore legs, rather upright on hind legs. Viewed from behind, her quarters and gaskins appear enormously large in proportion to the size of the animal. She is a wonderful jumper, and tried an eight-foot wall, but did not get over, having a log of wood tied to one of her hind legs. The eyes are large, quite black and very expressive. The muzzle is small and black, the nostrils large and open. The ears are long, outside light fawn colour, inside covered with long white hair. Outside the knees and hocks there are faint traces of three brown bars. The animal shows no indication of the cross, or shoulder stripe, found in other donkeys. She is at present over twelve hands high, but is not yet full grown.

"These animals have constantly been chivied on the Runn of Kutch for years past by parties of officers on horseback with spears; but, with the solitary exception which I have above mentioned, when a man named Elliott speared a jenny on the point of foaling, no wild donkey has ever been run down until my friend Nutt got hold of this one.

"This donkey was exhibited at the horse show in Poona, and was looked upon as the greatest curiosity and attraction there. She bites and kicks at everyone that approaches her but her own syce. It took a whole day to get her to stand steady, in order to take the photographs I send you; and at one time she lashed out with her hind legs, and kicked the photographer and his apparatus over. They say there is no possibility of ever taming her. FRASER S. HORE.

"Bombay, October 27, 1874."

These accounts of the untameable nature of the Onager and its extraordinary endurance appear to be based upon somewhat imperfect information. After their recent re-publication, I had the pleasure of receiving from Mr. J. L. Harrington, 14th Bombay Infantry, Assistant Superintendent of Police at Kathiawar, the following interesting account of the capture of several Onagers, which disproves the previously received information of their great speed, extraordinary endurance and extreme wildness. The statements made by Captain Nutt, in the *Oriental Sporting Magazine*, and Mr. Blanford in his "Fauna of British India," regarding the Onager, have unquestionably been greatly modified by the statements of Mr. J. L. Harrington, who writes as follows:

"Blanford, in his 'Fauna of British India,' states that there is no instance on record of wild asses being run down by a single horseman, and Mr. Tegetmeier also remarks that it is doubtful whether any Onager has ever been ridden down, except in cases of mares heavy in foal, and also states that even the young have only been captured by employing relays of horses.

"The above has been conclusively disproved by H.H. the Thakor of Morvi, whose State, which is in Kathiawar, is situated close to the Runn of Kutch, as the Onager has been ridden down and secured on several occasions during a period of three years, when riding them down was one of his highness's favourite amusements, undertaken chiefly to disprove the exaggerated opinion commonly held as regards their speed and endurance.

"On one occasion a band of eight wild donkeys were ridden down and secured on the east side of the Runn by a party of five riders, or, to be more exact, by a party of three, as the riding was really done by H.H. the Thakor Sahib and two of his riding boys. The riders averaged about 9 stone in weight, rode the same horses from start to finish, and kept together throughout the whole of the chase.

"As the above may not be considered a case of running them down by a single horseman, perhaps the following instance may suffice, viz., that on another occasion the Thakor Sahib and his two riding boys separated; the former succeeded in riding one down single-handed, and without change of mount, while the two boys secured another.

"The horses used in these rides were Walers, Arabs, and country breds, and in one ride where a wild donkey was secured, a 13·3 Arab pony was used. The fact may perhaps interest people that the country breds used were ordinary Kathiawar cobs about 14·1, and in the case of the Walers and Arabs used, no special selection was made of mounts, neither were the animals in special condition for the rides. The following facts will somewhat tend to discount the somewhat exaggerated ideas held concerning the speed and endurance of the Onager.

"The rides which ended in captures usually lasted about three hours; speed varied from a walk to a spurting gallop; the going was execrably bad, being chiefly ground covered at high tide by the sea, and consisted for the most part of mud, in which the horses sank fetlock deep, necessitating the greater portion of the chase being done at a walk. The distances covered in the different runs varied from twenty to twenty-five

Captured Onagers at Morvi.
(From a photograph.)

miles; no horse ever died during a chase or from the after effects.

"Although the Onager's speed is greater than that of a tame donkey, an ordinary 14.2 Arab can gallop them to a standstill, and the fact of the runs being so long was due more to the going than to any special endurance on the part of the wild asses. Practically, as regards endurance, they are as enduring as a horse in non-galloping condition, though the asses when caught, could hardly be called in galloping condition either.

"One of the most striking points in connection with these rides is the endurance shown by the horses used in capturing the wild asses; in fact, more wonderful than the endurance of the asses, who were on the ground they live on, whereas neither food nor water could be obtained for the horses, riders even having to carry their own drinking water. On one occasion horses were out without food or water from 7 a.m. one morning to 4 a.m. the next.

"Some twenty wild asses, big and small, were captured in these rides. When captured the wild asses were extremely vicious, bit and kicked, and it was found necessary to rope them before they could be led away. The statement that no varieties of the Asiatic wild ass have ever been domesticated would be deprived of some of its effect could your readers see the wild asses in the paddocks at Morvi.

"Though some of the captures remained excessively vicious, others became quite tame, and were ridden and driven just like tame donkeys. The young ones are as tame as dogs, and extremely fond of being fondled and played with.

"Those in the paddock at Morvi were exceedingly inquisitive, and had to be kept back while a photograph was taken, as nothing would satisfy them until they could sniff round the camera and see what the seemingly diabolical instrument was.

"A photograph, the only one of a batch of four taken which turned out passably (though the gentleman in the solah tope is meant to be a European), is herewith sent in proof of what may be done with them, and in it may be observed the inquisitiveness of the animals, a youngster, in his eagerness to find out what was going on, having come up behind and caused the syce on the right to move his hand. The treatment undergone by the donkey, on whose back a syce is seated, ought to be proof enough of her tameness, as her tail was held, not to keep her quiet, but to show what could be done to her. This particular donkey was ten months old when caught and frightfully wild; she is now about two and a half years old.*

"The engraving of the young Onager from the photograph forwarded by Mr. Fraser S. Hore is good with the exception of the legs and feet, which are made to look too coarse, the legs and feet of the wild ass being particularly clean, neat, and well formed.

"The same horses which were used to ride down the wild asses in the Runn have been used to ride down wolves and black buck (*Antelope bezoartica*).

"The information regarding the riding down of the wild asses on the Runn of Kutch was given to me by a well-known Kathiawar sportsman, the traffic manager of H.H. the Thakor Sahib of Morvi's State Railway, who was out with the Thakor Sahib on several occasions when they went after the Runn donkeys, whose riding weight, however, prevented him from being with the leaders in the runs when the Onagers were captured."

The supposed irreclaimable nature of the Onager is one of those fables that too often pass current in zoology. They descend from writer to writer, and are transmitted from one volume into others. Even in as recent a volume as "The Horse," by Sir W. H. Flower, we are told that the Asiatic wild asses outstrip the fleetest horse in speed and that none of them have ever been domesticated. Scores of Indian officers must have known that the Onager

* This photograph showing the docility of these animals is so conclusive in its evidence, that I have had it accurately reproduced, and have to express my deep obligation to Mr. J. L. Harrington for the kindness he has shown in forwarding it.

is readily domesticated, and that they occasionally become so tame as to be troublesome. Nevertheless, they are usually described as exceedingly vicious, although they are readily tamed, as stated by Mr. Harrington, and demonstrated by his photograph, and General Sir Samuel Browne informs me that these animals are generally to be seen at every station of the Punjab frontier force, from Kohat down to Rajanpore. He says that they are perfectly domesticated, and so tame that they find their way into the officers' houses and into the men's lines, and even come into the mess rooms and force their heads between the chairs to get bread from the table, and he instances one which was so civilised that it did not object to a little sip of pale ale. At various times General Sir Samuel Browne had no less than three with his regiment, and during the Mutiny one marched with the men from Peshawur as far as Lahore. She used to be amongst the officers' tents, roaming about the camp during the day, invariably moving on to the next encampment with the regiment. Another that was an equal favourite died from burns consequent on her tumbling into a smouldering heap. These animals, however, strenuously resisted being saddled, possibly from not having been broken-in when young; but one was known to Sir Samuel Browne as having been perfectly broken, and as being habitually ridden by a Belooch chief named Beeja Kham.

The Syrian variety of Asiatic wild ass, the Hemippe, is so closely allied to the Indian form as not to demand a distinct description.

THE KIANG.

(*Equus hemionus, var. Kiang.*)

ALTHOUGH regarded by the majority of naturalists as a local variety of the Asiatic wild ass (*E. hemionus*), the kiang, or dzeggetai, of Tibet differs so much from the better-known Indian wild ass as to render a detailed notice of it desirable. As will be seen by the engraving—which has been most accurately reproduced from a photograph of a kiang formerly existing in the Zoological Gardens—this animal differs from the onager, being larger and more powerful in the hindquarters, which appear abnormally developed in length and strength. It is also larger in size, reaching to 14 hands, and its colour is a rufous-bay, with a much narrower dorsal stripe than is found in the onager. Its voice is described as a neigh, and not like that of the onager—a shrieking bray.

The habits of the kiang are not as familiar to us as those of the wild asses of India, but they have been admirably described by more than one traveller who has visited the country. A very vivid sketch of the animal, from a sportsman's point of view, is to be found in Colonel Kinloch's "Large Game Shooting of Thibet and India," although the author's statements that there is a doubt as to whether it is a horse or an ass, and that it is more closely allied to the zebra, or quagga, than to the ass, will not be accepted by naturalists.

"The kyang (says Colonel Kinloch) prefers the most desolate places in the vicinity of lakes and large rivers. It delights in the coarse and wiry pasturage, its favourite food being a rough, yellow grass, as hard and sharp as a penknife.

"No animal is a greater nuisance to the sportsman. Very inquisitive by nature, as soon as kyang observe a strange

THE KIANG OF TIBET (*Equus hemionus*, var. *E. kiang*).

(From photograph of specimen in Zoological Gardens. 1859.)

object, they seem anxious to find out all about it; and often, when stalking, one is annoyed by a brute who snorts, cocks his ears, and then trots up to have a look at one. Any of his friends who may be near at once follow his example, more distant ones are attracted, and in a few minutes a herd of fifty or sixty may be galloping in circles, effectually alarming all the game in the country.

"They will also sometimes spoil sport by actually chasing and driving away other game from their pastures. I witnessed a case of this in the Indus valley in 1866, when some goa which I was stalking were hunted right away by some kyang. A friend of mine had his stalk at some antelope spoiled in a similar manner.

"In places where they have not been disturbed, kyang will frequently gaze at the sportsman within fifty yards without betraying any fear, but merely curiosity. On the more frequented routes which are annually traversed by tourists the kyang are much more shy, and seem to know the range of a rifle well. Of course, there is no sport in shooting such an animal; but the skin of one is occasionally useful to mend one's shoes with, and in some parts, as Ládák, the Tartars eat the flesh with avidity. I have tried it, and found it tough and coarse."

Colonel Kinloch adds that he saw it stated some years ago that a cross had been obtained between the kiang and the ass at the Jardin des Plantes, and that he should imagine that the cross between the kiang and the horse would be a most valuable animal, possessing all the good qualities of the ordinary mule, with greater size and strength, and better shape. I may state that in the opinion of experienced mule breeders, the points of the kiang are not such as would render its hybrid offspring as valuable as the ordinary mule.

I am not aware that the Tartars have ever utilised the kiang as a domesticated animal, and for any detailed description of its habits when in subjection to man we must turn to the very graphic account of the example formerly living in the Zoological Gardens, Regent's Park, which was written by Major W. E. Hay, who received it as a present from the Chinese Governor of Rûdôk, a hill fort situated beyond the Pâng Kông in Little Tibet.

Major Hay had endeavoured to procure two Tibetan dogs of enormous size, of the same breed that was described by Marco Polo as being of the size of donkeys. One of these, however, had died, and the person deputed, thinking Major Hay would prefer a kiang to a dog, secured the former. At that time it had never been haltered or handled. It was said to have been caught in a pit, and was much attached to a white Chûmûrti ghoont, which it would follow; but this animal being claimed by a Tibetan lama, Major Hay purchased a Tibetan mule to keep the kiang company. With this it did not agree, and the mule led anything but a happy life. The kiang would, however, follow it, and was always restless unless it had some equine animal in company.

It always showed the greatest aversion to pass over any insecure wooden bridges, and, when its companion had passed over a bridge, would wait until it saw that it had gained the opposite bank, and then would fearlessly plunge into the most rapid stream, and usually make a nearly straight course across. In leaving Kûllû for Simla it had to cross the River Biass, which was then a foaming torrent. It plunged in, but was carried down the stream several hundred yards, and landed upon an island, where it remained quietly until the following morning, when the mule was sent across to tempt it to follow to the shore, which it did. The Sutlej was at this season so full, and running at such a frightful pace, that Major Hay deemed it advisable to throw the animal and secure it upon a raft, which was with great difficulty got across. It was at Simla during the whole of one rainy season, and did well, although Adolph Schlagintweit had given it as his opinion that the animal could not live under an elevation of 10,000ft. above the level of the sea. It was then marched to Ferozepore. On reaching the plains it seemed rather inclined to enjoy freedom, and occasionally required four men to hold and lead it, and even then on several occasions it got away, but was not very difficult to secure again.

At Ferozepore the mule which had accompanied it was dismissed, and the kiang taken to Kurrachi by water, in a boat purposely fitted up. There was much difficulty in getting it on board. It was disembarked at Kothree, and marched across the country to Kurrachi.

After keeping it a month at Kurrachi, it was shipped in the barque *Sumner*, a large quantity of hay, kirbee, dried lucerne, and grain, being provided for it. The latter was worm-eaten, and it was long before the animal could be induced to touch it. The passage was very long, and, provisions running short, the kiang was twice reduced to eat the straw with which the sailors' beddings had been stuffed.

At first it refused to drink any tainted water, but, before reaching St. Helena, where fresh supplies were obtained, it would eat or drink almost anything. On board ship it became exceedingly knowing, and balanced itself so beautifully that it was not slung, unless the weather was very rough. In an actual gale the poor creature laboured dreadfully, and seemed grateful for attention. It became latterly extremely docile, and always knew ts owner by his voice. In crossing the line the weather was very trying, and the kiang suffered greatly from the extreme heat. With the exception of about three days, it always had a voracious appetite, and consumed in four as much food as had been laid in for six months.

Major Hay states that in Tibet the kiang breeds with the horse, and that their produce is highly valued; and he adds, although not on his own knowledge, that the hybrids are regarded as fertile, which is in the highest degree improbable. Of its voice Major Hay says:

"I have often heard this one attempt a neigh, but it is a sad failure; at the same time it as little resembles the bray of an ass; indeed, its mode of calling to its companion is, like itself, quite unique. I feel quite confident that this female kiang may be got to breed with a horse. I always found the kiang very susceptible of kindness, its satisfaction being usually expressed by throwing its ears forward; it generally shows a sort of pettish displeasure when anyone is leaving it after giving it bread, &c. I twice placed a native of India on its back, but this was after it had gone a march, when it was slightly distressed by the heat of the weather; it took no notice whatever of its rider. I was convinced of the kiang's specific difference from the wild ass of Scinde when I saw one of the latter at Delhi, intended for conveyance to England, and this made me persevere the more to get it home. I have often watched the herds of this animal on the plains or slopes of hills in Tibet; one invariably stands sentry at from 100 to 200 yards from the flock, and when danger is at hand he commences walking leisurely towards his companions. They take the alarm, and, as soon as he comes up, off all go together in a trot or canter, as the case may require. I don't know to what space to limit the range of the kiang. Marco Polo speaks of asses, but evidently alludes to those of Persia. Huc and Gabet evidently saw them towards Lassa; and I have been told that they are to be met with on all the level country between Ladak and Lassa, or in the valleys between the various ranges. I have seen them only north of the great Himalayan ranges; first upon the Rupfcher plains and in the neighbourhood of the Salt Lakes, often in company with the Ovis ammon or 'nyan.' I have also seen them north of the Pang-Kông lake. The passes from Hindustan into Tibet are never open before June, when I have seen flocks of the kiang feeding almost entirely on the roots of a species of artemisia, or wormwood. Their natural enemies besides man seemed to be a panther, which lurks amongst the rocks, and a large species of wolf. I have found their skeletons on the melting of the snow. Beyond the Pâng-Kông lake I was informed that in winter many of them were to be seen in the Shap-Yok valley, in company with wild yâks or dông, also the 'nyan' (*Ovis ammon*) and the 'sûs,' or Tibetan antelope (*Panthalops hodgsoni*). A few tamarisk bushes seem then to support them, and at the end of winter all these animals are spoken of as being like walking skeletons. I have sometimes approached flocks of kiang quite close, at other times could not get within a mile of them. On one occasion two kiangs followed a pony on which I had a servant mounted; in fact, kept so close that my servant feared they were going to attack him. I never could ascertain satisfactorily when the kiang breeds; but I think it must bring forth in winter, for I have seen a mare shot with a young one in the womb, nearly mature, in August; and in the many flocks I have met with running wild I never perceived a foal that I should have taken to be of less than six months old. When very young the hair of the foal has the appearance of wool. The winter coat of the adult is also very thick and curly, and is of darker colour than its summer coat. It appears to shed its winter coat in May. The kiang may be said to inhabit plains and undulating hills, at from 15,000ft. to 16,500ft. above the sea; if found in the steeper hills they have been driven there. It is most wonderful to see the rapidity with which they can ascend mountains, and although they descend quickly I never saw one lose its footing. After they have been pursued for some time on the hills and driven on to the plains, they will frequently make a charge past you at about 100 yards distance in preference to ascending the steep parts again, thus showing their preference for level ground. They are almost always seen in the neighbourhood of lakes or ponds in the unfrequented spots which are usually beyond the sportsman's beat."

I have drawn largely on this exceedingly interesting account of the habits of the kiang, as I am not aware that it has been republished since its appearance in the "Proceedings of the Zoological Society" more than thirty years since, and is not likely to have come under the notice of the general reader.

THE MOUNTAIN ZEBRA (*Equus zebra*).

CHAPTER VI.

THE MOUNTAIN ZEBRA.

(*Equus zebra. Linnæus.*)

THE species of the Equidæ distinguished by their bodies being marked by stripes are restricted in their geographical range to the African Continent. They were formerly, by some naturalists, regarded as constituting a distinct genus (*Hippotigris*), differentiated from the other asses by their stripes; but, as is generally recognised, mere variations in colour and markings do not constitute good generic differences, and the zebras are now regarded as constituting one genus (Equus) with the horses and asses. The fact that all species of this group are occasionally more or less marked with stripes is in itself a fact opposed to their presence being regarded as a good generic distinction.

The mountain zebra—the Wilde Paard, or wild horse of the old Dutch African colonists—was the one which was first made known to Europeans, and, being formerly abundant in the mountainous parts of the Cape Colony, was called the common zebra; but now, owing to the advance of the colonists, it has become rare, and during the whole time that the Zoological Society has been in existence it has received but three specimens, one of which was acquired in 1864, being presented by Sir P. Woodhouse, the Governor of the Colony, the second purchased a few years later, and in 1887 a young male was obtained, whose portrait, taken at the time, is given in the plate.

The mountain zebra in form more nearly resembles the ass than the one now better known as Burchell's zebra. It is also the smaller of the two, being about 4ft. across the withers. It has longer ears than the Burchell's zebra, and a considerably shorter mane. The general ground colour is white, but the stripes are black, and broader than the intervals that separate them. The stripes on the body are all nearly perpendicular. The muzzle is a bright brown. If we except the abdomen, which has a longitudinal stripe along it, and the inside of the thighs, the whole of the body is striped, the legs being covered with transverse bands reaching down to the hoofs, and the base of the tail itself is transversely marked. It is remarkably distinguished from the other striped equine animals by what has been termed by some travellers the "gridiron" marking above the tail, formed by a series of short transverse bands passing from the middle dorsal stripe outwards, and generally joining the uppermost of the broad stripes on the haunch. This is always present, and serves to distinguish at once the mountain zebra, from the other striped members of the group. It is also characterised by the presence of a distinct, though small, dewlap, which is well shown in the vignette at the end of this chapter.

It is interesting to note that the gestation of the zebra approaches to that of the ass rather than that of the horse. The Earl of Derby, writing of one in the "Knowsley Menagerie," said, "Mine has gone more than a week over twelve months."

The employment of arms of precision has already effected a great change in the fauna of South Africa, and to learn the habits of the mountain zebra it is necessary to turn to older writers, for it is now so scarce that it has even been supposed to be extinct in the district. Fortunately we have very satisfactory accounts of it in comparatively recent writers. In the magnificent folio on the game and wild animals of Southern Africa, published by Capt. W. Cornwallis Harris in 1840, a very full description of this animal is given. Of its habits Capt. Harris writes as follows:

"Restricted to the mountainous districts of Africa, from Abyssinia to the southernmost portions of the Cape of Good Hope, this beautiful and wary animal never of its own free will descends into the plains, as erroneously asserted by naturalists, and it therefore never herds with either of its congeners, the quagga and Burchell's zebra, whose habitat is equally limited to the open and level lowland. Seeking the wildest and most sequestered spots, the haughty troops are exceedingly difficult to approach, as well on account of their watchful habits and extreme agility and fleetness of foot as from the abrupt and inaccessible nature of their highland abode. Under the special charge of a sentinel, so posted on some adjacent crag as to command a view of every avenue of approach, the chequered herd whom 'painted skins adorn,' is to be viewed perambulating some rocky ledge, on which the rifle ball alone can reach them. No sooner has the note of alarm been sounded by the vidette, than, pricking their long ears, the whole flock hurry forward to ascertain the nature of the approaching danger, and, having gazed a moment at the advancing hunter, whisking their brindled tails aloft, helter skelter away they thunder, down craggy precipices and over yawning ravines, where no less agile foot could dare to follow them."

Burchell, who was well acquainted with both this and the other species which was named after him, calls attention to the constricted character of its hoofs, which are adapted to rocky mountainous regions, those of the E. burchellii being fitted for the plains.

Although the true zebra is much more beautiful in its markings than the allied species known as Burchell's zebra, there can be no doubt that it is the more asinine in its formation of the two, not only in the form of the head and tail, but most markedly in the length of the ears; nevertheless, the animal is full of grace and beauty. It is true its shoulder is straighter than would be approved in a horse, that the quarters are shorter, the neck thicker, and the cannon bones longer, but no one can look at the animal without being struck with its extreme beauty.

From its smaller size, straighter shoulders, and more asinine form, the mountain zebra is less adapted for the service of man as a domestic beast of burden or draught than the Burchell's zebra; nevertheless, it can be tamed and ridden, and Captain Hayes has most obligingly allowed me to use a photograph, from which the accompanying illustration was taken, showing Mrs. Hayes riding one of these animals that had been some time in captivity in a travelling menagerie in India. He informs me, however, that it is a much more difficult animal to handle and break in than the comparatively stronger and larger Burchell's zebra.

In his recent work on the "Points of the Horse," Captain Hayes, speaking of this zebra says, it has a thicker neck, and its legs, especially as regards the back tendons and suspensory ligaments, are not so well suited to civilized requirements as those of Burchell's zebra. At present it is met with in a wild state only on a few mountain ranges of the southern part of Cape Colony, where it is preserved. There is a herd on a farm near Craddock, a small town in the eastern province; it is much wilder and more intractable to handle than the Burchell zebra. The height of the mountain zebra he gives when fully grown as twelve hands. With regard to its utilization as a domestic animal, Capt. Hayes says that he has been informed that it has been successfully "inspanned" in South Africa, but that he has never heard of its being put into draught between the shafts, and he points out that

EQUUS ZEBRA BROKEN TO SADDLE.
(From a photograph by Capt. M. H. Hayes.)

the steadiness of an animal is much more accurately tested by having to bear a weight upon its back than by merely pulling against a collar when in a "span," and still more so by carrying a rider than when in any kind of harness. After making every inquiry in the colony he was unable to obtain a single authenticated instance of any person ever having ridden a mountain zebra, and he is justifiably proud of the fact that in two days he broke in the old stallion, shown in the engraving, to be sufficiently quiet to permit Mrs. Hayes to ride, and to be photographed whilst on his back.

CHAPTER VII.

GREVY'S ZEBRA.

(*Equus grevyi.*)

OUR first knowledge of this animal dates from 1882. On December 19th in that year Mr. P. L. Sclater exhibited, at the meeting of the Zoological Society, photographs of a zebra, recently living in the Jardin des Plantes, Paris, which he had received from M. A. Milne-Edwards, and he pointed out the differences that he considered separated this animal from the common or mountain zebra. At that time a single specimen of the species had been sent alive by King Menelek of Shoa to the President of the French Republic, but it unfortunately died after a short residence in the Zoological Gardens at Paris.

This animal is doubtless identical with that common in Somali-land, described by Dr. Emin Bey as existing in large numbers in Lattako. This naturalist, however, identified it with the ordinary Equus zebra.

Eight years afterwards, that is in 1890, Mr. Sclater exhibited, at the Zoological Society, a skin of this zebra, which was received from northern Somali-land, and said:

"I have recently again examined the typical example of this species, now mounted in the new gallery of the Jardin des Plantes, and am still more confident of its distinctness, as shown by the narrowness of the black stripes, the difference of the markings, and the white spaces on the forehead and on each side of the dorsal stripe in the northern species.

"Being anxious to know whether the Berg-zebra of Somali-land, spoken of by Herr Menges ('Zool. Garten,' 1887, p. 263) as found in the mountains of that country as far north as 8° North Latitude, belongs to the E. grevyi, I requested Mr. Hagenbeck to endeavour to obtain for me a skin of this animal. This he has most kindly done through the intervention of Herr Menges."

SKIN OF EQUUS GREVYI FROM SOMALI-LAND.

That there is considerable variation in the markings of the species is evident from a comparison of the engraving of this skin with that of the animal received from Shoa. The croup of the latter, which lived for a short time in the Jardin des Plantes, as shown by the engraving, is white, whereas in the Somali skin it is covered closely with small stripes, which bear a distinct relation to those which constitute the "gridiron" markings in the true Equus zebra.

Sir William Flower, writing of this species in "The Horse," says :

"Being obviously different from any that had hitherto been seen in Europe, it was named by M. Milne-Edwards Equus grevyi, in compliment to his political chief. On a white ground colour it is very finely marked all over with numerous delicate, intensely black stripes, arranged in a pattern quite different from those of the other species. In view of the great variability of the markings of these animals, as long as but one individual of this form was known, some doubts were expressed as to whether it might not be an exceptionally-coloured individual of one of the other species; but, subsequently, other specimens, presenting almost exactly the same characters, have been received from Somali-land, and it seems probable that all the zebras which we know to exist in the northern districts of East Africa belong to this species."

And writing in "Mammals Living and Extinct," the same author states that "In many of its characters it resembles E. zebra, but the stripes are much finer and more numerous than in the typical examples of that species, and it has a strong black and isolated dorsal stripe."

The publication of the account of Grevy's zebra in the *Field* elicited the following letter, containing interesting particulars respecting the distribution and habits of the animal, from Capt. H. G. C. Swayne, which was dated Aden, July 20th, 1893:

"While returning from an expedition in Northern Somali-land, I received the interesting notes on 'Wild Horses, Asses, and Zebras,' in which remarks were made concerning the new zebra. Equus grevyi. Most of the skins which have hitherto reached the Somali coast have been brought down by natives, and as I believe these zebras have been shot now for almost the first time by Europeans, a note on their habits may be of interest.

"I found they did not range further north on my route than about 7° 50′ of latitude, and thence to the Webbe Shabeli river at Imé, on the Galla border, they were common. The zebras, of which I saw several herds at different times, were always found on low plateaux covered with scattered or thick thorn bush and tall, feathery 'durr' grass, with red gravelly soil and rocks cropping up now and then. I saw none of their tracks in the wide open grass plains, though this was not, I believe, the experience of another sportsman whose route lay about 100 miles to the eastward of, and parallel to, mine. The zebras, when I saw them, were in herds of under a dozen, and they were so tame that it was only because I had a large following to feed that I was induced to shoot them. I have several skins, and the stripes of adult ones only approach 'intense black' over the withers; elsewhere they are of a very deep chocolate colour, changing to light tan on the forehead and muzzle.

"In the skin of a quite young zebra which some natives brought me, the stripes were light brown, except on the withers. I notice that skins brought down by natives and sold in Aden seem to fade, and appear nearly dull black. The stripes on all the skins of some 200 zebras which I saw alive, at one time and another, were of the same narrow type on the flanks, showing no variation in pattern so far as I could see."

At a later date Captain Swayne, in his valuable field notes on the Game Animals of Somali-land, published in the "Proceedings of the Zoological Society" for 1894, writes of this animal as follows:

"Grevy's Zebra (the Somali name of which is *fer'o*) was, I think, first shot in Somali-land by Colonel Paget and myself on our simultaneous expeditions last spring.

"I found them first at Durhi, in Central Ogaden, between the Tug Fafan and the Webbe, about 300 miles inland from Berbera. I shot seven specimens, all of which were eaten by myself and my thirty followers; in fact, for many days we had no other food; and this was no hardship whatever, as the meat is better than that of many of the antelopes. The flesh is highly prized by the Rer Amaden and Malingur tribes.

"The zebra was very common in the territory of these two tribes. The country there is covered with scattered bush over its entire surface, and is strong and much broken up by ravines; the general elevation is about 2500ft. above sea-level.

"The zebras, of which I saw probably not more than 200 in all, were met with in small droves of about half a dozen on low plateaux covered with scattered thorn bush and glades of 'durr' grass, the soil being powdery and red in colour with an occasional outcrop of rocks. In this sort of country they are very easy to stalk, and I should never have fired at them for sport alone. I saw none in the open flats of the Webbe valley, and they never come near so far north as the open grass plains of the Hand; Durhi, south of the Fafan, being their northern limit.

"The young zebras have longer hair, and the stripes are rather light brown, turning to a deep chocolate, which is nearly black in adult animals.

"After firing at one of a drove of zebras, I was sorry to find on going up to it that it was a female, and that its foal was standing by the body, refusing to run away, though the rest had all gone. We crept up to within ten yards of it, and made an unsuccessful attempt to noose it with a rope weighted by bullets, but it made off after the first try. We must have been quite five minutes standing within ten yards in the thick bush while we were preparing the noose.

"Zebras are very inquisitive; when I was encamped for some days at Eil-Fúd, in the Rer Amaden country, the zebras used to come at night and bray and stamp round our camp, and were answered by my Abyssinian mule. The sounds of the two animals are very similar."

The late Mr. J. T. Tristram-Valentine wrote:

"I have read Mr. Tegetmeier's remarks on Equus grevyi, which appeared in the *Field*, with great interest, and I should like to say a few words on the subject of the Somali-land zebra. Some time since I exhibited, at a meeting of the Linnean Society, a flat skin of this zebra, which I had received from Captain H. D. Merewether, then Assistant Political Resident at Berbera, and pointed out that it differed from the type specimen of Equus grevyi in that the stripes were brown (red-brown) upon a pale sandy or rufescent ground, instead of black on a white ground; and I suggested that this might be the desert form, the type specimen representing the mountain form. I have since seen several skins of this zebra, and they all of them agree in colouration with my own, as, indeed, does the one in the British Museum, which was exhibited by Dr. Sclater at a meeting of the Zoological Society, and figured in the Proceedings. And I am told by Captain Merewether that, though he had seen dozens of them at Berbera, brought by caravans from the Dolbahanta country, he has never seen one with black stripes. I may add that there is every reason to believe that the skin obtained by Dr. Sclater, though said to have been received from 'northern Somali-land,' was brought to Berbera by one of the Dolbahanta caravans. I may further add that the description of the ground on which Colonel Paget is said to have found these animals—flat ground, in open scrub, about 150ft. above the level of the river—exactly agrees with the description given me, which, in conjunction with the coloration of the animal, caused me to suggest that this was a desert form. In conclusion, I would observe that the country from which the Somali-land zebras have been procured is some hundreds of miles distant from the mountainous territory of Shoa, from which the type specimen of Equus grevyi was obtained."

The most recent information we have regarding this animal is in a letter from Mr. A. H. Neumann, Laiju, East Central Africa, April 16, 1894, who writes:

"As we emerged from the bush we saw zebra ahead of us. . . . I soon saw that they were not the common Burchell's by their great wide ears and different markings. . . . I gave one a shot, and following, found him lying down as if alive, but really dead.

"A beautiful creature he was—a fine young stallion, larger and far handsomer than Burchell's zebra, the stripes much narrower, except one very broad dark one down the back, with wide light ground on either side. The cry of this zebra is quite different from the bark of the commoner kind, being a very hoarse kind of grunt, varied by something approaching a whistle. The Mackenzie river seems to be about their limit here, as on this (west) side of its most easterly branch I saw only Burchell's."

The old doctrine of the immutability of species and their separate and distinct creation is one that is not now held by the majority of modern naturalists. Disputing, therefore, as to whether two closely allied animals are specifically or sub-specifically distinct is almost a waste of words. We know that a species spread over a wide area will change according to the conditions of life, until at last the two extremes are so diverse as to be regarded as distinct species, but no one can say where one species ends and the other begins, for they merge gradually into each other.

If I might be bold enough to express an opinion, I would say that Equus grevyi and Equus zebra are the same animal modified slightly by a long residence, possibly for many thousands of years, in different localities. The skin which Mr. Sclater reproduces as evidence of their distinction, appears to me the most convincing proof of their identity. There are to be seen in both animals the same transverse bands on the legs, the same general disposition of the stripes on the body, and on the neck. There are even in the E. grevyi the rudiments of the gridiron marks on the hind quarters of the E. zebra, and almost the only difference is the larger number and smaller width of the stripes in E. grevyi. The "strong black and isolated dorsal stripe" in the type specimen at Paris, on which Sir William Flower places so much reliance, is merged in the transverse stripes in the Somali-land skin. I am free to confess that I think the specific distinctions that are made to depend on alterations and locations of colour are extremely unsatisfactory. That the narrowness or broadness of the stripes, or their being more or less numerous, should cause animals to be regarded as being specifically distinct, appears to me to be a fallacious idea.

BURCHELL'S ZEBRA (*Equus burchellii*).
(From specimen in Zoological Gardens, 1887.)

CHAPTER VIII.

BURCHELL'S ZEBRA.

(*Equus burchellii. Gray.*)

THE best known zebra at the present time is that which was named after Burchell, the African traveller. The species is still common in some parts of South Africa, and is now being utilised in the coach teams in the Transvaal. The Burchell differs from the mountain zebra (*E. zebra*) in several essential parts. It is a larger and stronger animal, with shorter ears, which are rarely more than 6½in. in length, and have a much larger proportion of white, a longer mane, and a fuller and more horse-like tail. The general colour is pale yellowish brown, the stripes being dark brown or nearly black. There is usually a longitudinal stripe along the under side. The dorsal stripe is defined by a white line over the haunches, and there are not any stripes proceeding from it at right angles as in the mountain zebra.

There are two, if not more, well-marked varieties of Burchell's zebra. The one originally described was remarkable for the absence of markings on the fore legs and on the tail. In the other variety the limbs are covered more or less completely with transverse stripes, and this form has been named after its first discoverer, Mr. E. Chapman. Chapman's zebra was originally described by Mr. E. L. Layard in the Zoological Society's Proceedings for 1865.

Mr. Layard compared it with the mountain zebra, and said it was distinguished by the union of all the black stripes, with a median one on the belly.

"The new animal (he wrote) also differs from the other zebras in having the callosities on the legs far larger and of a more rounded shape, in having shorter and more equine ears, measuring only 6½in. instead of 11½in., and in having a shorter and more equine head and tail. The hoofs are also flatter than in the common zebra, and not adapted for mountain work. The mane grows several inches down on the forehead, and stands up between the ears, so that when seen in full face it stands far higher than them. They roam in large herds, and are first met with about 200 miles from the coast inwards on leaving Walwich Bay, where Equus zebra (or, rather, a variety of that animal) prevails. The height of a young male shot in 1862, at the shoulder, was 4½ft.; at the rump, 5ft."

This animal was also described in the same communication by Mr. Baines, and figured by Mr. Wolf. The distinction between Burchell's and Chapman's zebras—if the latter is allowed to stand as a true species, which is very doubtful—is merely that of marking, and it has not affected apparently the character of the animal. This subspecies, E. chapmani, has a very wide range. The skin which is represented in the engraving was forwarded to Mr. P. L. Sclater (who exhibited it at the meeting of the Zoological Society) from Masailand, East Africa, which is between the Lake Victoria Nyanza and the east coast.

Burchell's zebra is not only a larger, but, from a utilitarian point of view, a much better-formed animal than the mountain zebra, which may be described as far more asinine in form. It is also more easily broken to harness, and readily becomes a domesticated animal. Some years since I visited Theobalds, the seat of Sir H. Meux, to see a mare of this species that had several hybrid foals, and, although she had never been handled, I walked quietly up to her in her paddock and placed my hand on her withers without her evincing any uneasiness—in fact, she was much more docile than her hybrid offspring.

It is only recently that the Cape Colonists have arrived at the conclusion that Burchell's zebra is a desirable beast

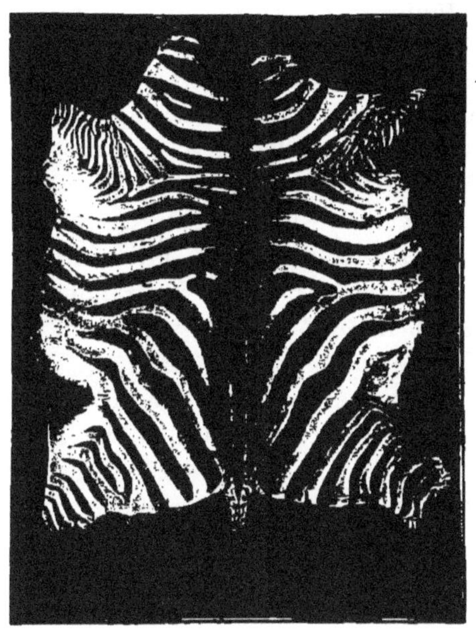

SKIN OF BURCHELL'S ZEBRA FROM MASAILAND, EAST AFRICA.

of draught and of burden. This fact, however, may be regarded as having been very distinctly demonstrated. A number are now being driven, and I reproduce a photograph of a team of Burchells driven four-in-hand in a two-wheeled Cape cart. This demonstrates the fact that they can not only be employed in teams with other animals, but that they are sufficiently docile to be used alone. In 1893 Burchell's zebras were on sale in the Cape at prices varying from £10 and upwards, and several have recently been imported into this country for the purpose of demonstrating their utility as beasts of draught, to ascertain their prolificacy in this country with their own and other species, and their capabilities of adaptation to the conditions of life that here obtain.

Writing of this species from Johannesburg in the Transvaal in December, 1892, Mr. Harold Stephens states:—

"You will be pleased to hear that an effort is being made in the Transvaal to domesticate and use the zebra for purposes of draught. On hearing that Messrs. Zeedesberg, the coach contractors, who run passengers and mails from Pretoria in the Transvaal to Fort Tuli in Mashonaland, had been successful in their efforts in training the zebra, I determined to make full inquiries when next in Pretoria.

"Pretoria, the capital of the Transvaal, is a very pretty little town situated about thirty-five miles to the north of Johannesburg, and as the sittings of the High Court are held there, it was not long before I found myself, in company with others, journeying towards it in a coach and ten horses, the usual method of travelling out here. Mr. James Zeedesberg, who I met by appointment the next afternoon, told me that his firm about two months ago bought eight half-grown wild zebras from a hunter named Groblaar. Groblaar caught them in a wild state between four and five months ago by riding after and lassoing them. During the last month they have been in training for harness, with the result that four of them are perfectly quiet and well trained, and the remaining four partially trained. The place where they are located is at the station in Petersberg, in the district of Zoutpansberg, Transvaal. It appears

Utilisation of Burchell's Zebra in the Transvaal.

they are a little timid at first when the harness is being put on; but afterwards they are all right, and Mr. Zeedesberg believes in a month or two's time they will be as steady as horses. They pull well and are very willing, and never jib—a vice which is very prevalent in the horses of this country. In fact, one of them will do his best to pull the whole coach himself.

"As you will see by the photograph which I send you, they are now being used in one of Messrs. Zeedesberg's coaches; and Mr. James Zeedesberg says they are so satisfied with the experiment, so far as it has gone, that he is going to extend it, with the object of ultimately substituting them for mules, as the zebra is free from that scourge of South Africa commonly called "horse sickness," which any of your readers who have been out here will know costs an enormous amount to coach proprietors in horse flesh during the summer season. In some parts of the low country it is quite sufficient for a horse to be left out all night in the veldt (grass) to ensure its death from this dreaded disease.

"The zebras, when inspanned (harnessed to the coach), stand quite still and wait for the word to go, they pull up when required, and are perfectly amenable to the bridle, and are softer mouthed than the mule. They never kick, and the only thing in the shape of vice which they manifest is that, when first handled, they have an inclination to bite, but as soon as they get to understand that there is no intention to hurt them they give this up. Four of these zebras are now inspanned and driven in a team together, and are as reliable and good as the best mules; the other four, being older, require a little more time to get them perfectly trained. The illustration shows four zebras inspanned with mules in one of the coaches at Petersberg.

"The intention is to buy more and run them regularly in the up-country coaches from and to Mashonaland, and this will not be done as a useless experiment, but with a practical object, and if it succeeds, as Mr. Zeedesberg believes, it will be the means of saving them hundreds of pounds, which they now lose annually through horse sickness. Later on attempts will be made to cross them with the horse, with the object of getting a larger and handsomer mule than the ordinary cross with the donkey, and probably superior in every way.

"It will be interesting to watch the progress of these experiments, which may bring about a new and important industry, for if the cross between the zebra and the horse can be brought about without difficulty, it will not be long before these animals will be preferred to ordinary mules, numbers of which are shipped out here from Monte Video, while those who are interested in natural history will only be too pleased at the chance of adding the zebra to the list of our few domesticated animals."

In reference to this interesting letter, Capt. M. H. Hayes writes: "The zebra referred to by Mr. Harold Stephens is the Equus burchellii, a very easy animal to tame. At the Agricultural Show which was held at Pretoria, April, 1892, I broke in a Burchell's zebra, which belonged to Mr. Ziervogel, quiet to ride after about half an hour's handling, without having to throw him down, tie him head to tail, or to resort to any of the other heroic methods of the horse-tamer. Equus zebra is of quite a different temper, and is an extremely difficult animal to subdue. I look forward to the Burchell's zebra becoming a very useful domestic animal; but the conformation of Equus zebra is not suited to civilised requirements." And in his valuable work,

Burchell's Zebra in Cape Cart.

recently published, on "The Points of the Horse," treating of this species, he writes:

"Its legs, below the knees and hocks, from their 'flatness,' with the back tendons and suspensory ligaments clearly showing, are much more like those of a well-bred horse than are those of the mountain zebra. It further resembles the horse by having a fairly lissom neck and a well-rounded barrel, and in the size of its head and ears. The typical Burchell's zebra has no dark stripes, or only very slight ones, below the elbows and stifles, on the legs. The Orange River has been generally regarded as its southern limit. Mr. F. C. Selous, the celebrated African sportsman and naturalist, tells me that it 'was first discovered by Burchell near the Orange River in Southern Bechuanaland. It is still to be met with in Kama's country, and along the northern and eastern borders of the Transvaal In the neighbourhood of the Pungwe River it exists in very great numbers, herds of hundreds together being common.' It is probably widely distributed throughout Central and Eastern Africa. On account of the fact that this zebra, when in a wild state, possesses immunity from the effects of the bite of the tsetse fly, which is certain death to horses, I strongly advocated, while I was in South Africa, the taming and employment for harness or saddle of these animals in 'fly' infected districts. With respect to this subject, Mr. Selous writes to me that: 'Although Burchell's zebra, born and brought up in the 'fly' country, does not suffer from its bite, it is my opinion that if a young one was caught and brought up in a locality where there was no 'fly,' and was then taken into a 'fly' infested district, it would die. This, however, is only my opinion.' As the Burchell zebra is comparatively easy to break in, and as it will breed in confinement, there is but little doubt that it will in time become domesticated. If, as is quite possible, it possesses little or no tendency to contract 'horse sickness' it will prove a valuable means of conveyance in South Africa."

The advantages of the utilisation of Burchell's zebra as

a beast of transport are so evident that they have commended themselves to all military officers familiar with African animals. Captain Lugard, in his work on our East African Empire, after speaking of the elephant and other beasts of transport, writes as follows :—

"There is another animal in East Africa which offers, as I have said, possibilities of domestication, viz., the zebra. If this animal were tamed, the question of transport would be solved. Impervious to the tsetse-fly, and to climatic diseases, it would be beyond calculation valuable.

"The species found both in East Africa and Nyasaland is 'Burchell's' (*Equus burchellii*). It is a lovely animal, of perfect symmetry, and very strongly built, standing about fourteen hands high. The bright black and white stripes of the zebra would appear to be the most conspicuous marking imaginable. Yet, when standing in the sparse tree-forest, it is one of the hardest of all animals to see, and even after it has been pointed out to me close in front, I have sometimes been unable to distinguish it, though, as a rule, I am even quicker at sighting game than a native. The flickering lights in a forest, and the glancing sunbeams and shadows, are counterfeited exactly by the zebra's stripes, and thus it is that nature affords protection to an animal otherwise peculiarly liable to destruction in the jungle; in the open plains, where his enemies cannot steal upon him unawares, he can rely for his safety on his own fleetness.

"The zebra throughout East Africa, so far as my observation goes, has suffered complete immunity from the cattle-plague, which has attacked most of the rest of the game. This disease has now spread south to Nyasaland, and Mr. Sharpe reports that between Mweru and Tanganyika Lakes he saw numbers of dead zebra. Mr. Crawshay also reports great mortality among the zebra in that district. Here—in Masailand and on the Athi plains—herds numbering their thousands may be seen, and these have not suffered from the plague.

"Some years ago (1888) I advocated experiments in taming

Burchell's Zebra.

Mountain Zebra.

Burchell's and Mountain Zebras contrasted, showing equine and asinine character respectively.

the zebra, and I especially suggested that an attempt should be made to obtain zebra-mules by horse or donkey mares. Such mules, I believe, would be found to be excessively hardy, and impervious to the fly and to climatic diseases.

"When we recollect that the zebra is found all the way from the coast to the far parts of Uganda (I have seen them in Buddu), and that countless thousands roam on the level plains of Masailand, where every possible facility is afforded by the open nature of the ground either for riding them down and lassoing them, or for capturing them by driving them into kraals or kheddahs, we shall realise that, when once the possibility of training the zebra as a pack or draught animal is demonstrated, the question for animal transport for East Africa is finally solved. The elephant would be invaluable in many ways, but his utility as an agent for the development of the country cannot be compared with that of the domesticated zebra. I would even go further, and say that their export might prove one of the sources of wealth and revenue in the future, for, as everyone knows, the paucity of mules, both for mountain batteries and for transport purposes, has long been one of the gravest difficulties in our otherwise almost perfect Indian army corps. I would therefore advocate that the zebra should be at once protected, and its slaughter absolutely prohibited. Its capture might be made a State monopoly."

The ready training of Burchell's zebra as a draught animal has been demonstrated by the Honourable Walter Rothschild, who has placed three in the hands of a very careful breaker, and they are now being driven both in single and double harness in the streets of London.

I have to express my thanks to Capt. Hayes, and the London Stereoscopic Company, for the permission to copy the admirable photograph of Burchell's zebra which illustrates this chapter, and to the Zoological Society for the permission to use the two engravings which demonstrate so convincingly the horse-like form of Burchell's zebra, and its fitness for the service of man as compared with the more asinine conformation of the mountain zebra. At the Jardin des Plantes experiments are now being made on the production and utilisation of cross-bred animals between the mountain and Burchell's zebras.

Burchell's zebra is now often called the quagga in some districts of South Africa—an error which has unfortunately been followed in the late Lord Randolph Churchill's work on "Men, Mines, and Animals in South Africa."

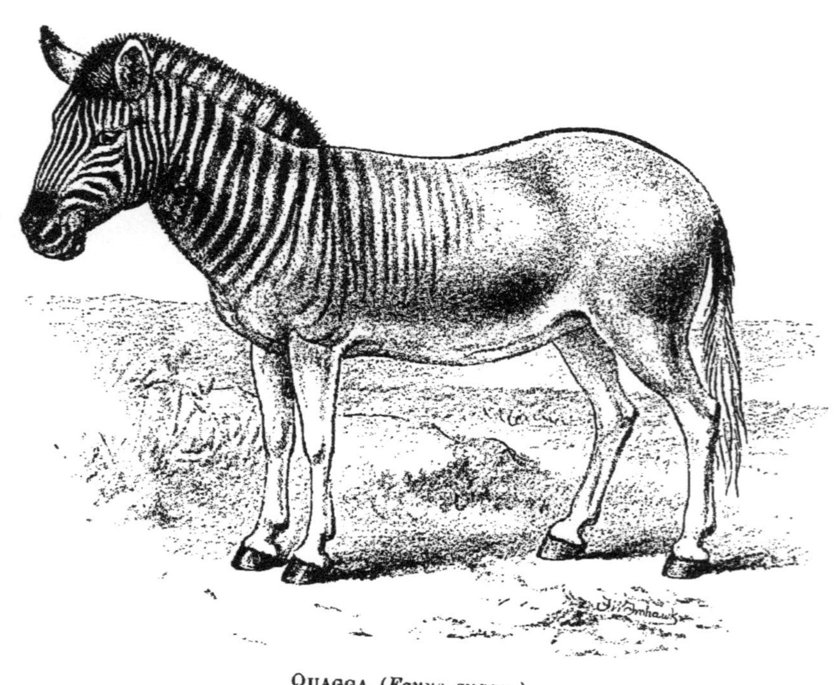

Quagga (*Equus quagga*).

CHAPTER IX.

THE QUAGGA.

(*Equus quagga. Linnæus.*)

The quagga, the last remaining species of the Equidæ that I have to describe, is probably at the present time an extinct animal, although within my own knowledge specimens existed in the gardens of the Zoological Society, and its hybrids, bred in the gardens, were driven about London in a light tandem, which was employed to convey vegetables from Covent Garden Market to the Regent's Park gardens. Before the foundation of the society, a pair of imported quaggas were in the early part of the present century driven about London in a phaeton by Mr. Sheriff Parkins, and Lieut.-Col. C. Hamilton Smith, in his unpublished volume on the Equidæ, 1841, states that he drove one in a gig, and that its mouth was as delicate as that of a horse; he further stated that it had better quarters and was more horse-like even than Burchell's zebra, and added: "It is unquestionably the best calculated for domestication both as regards strength and docility," and he gives drawings taken by his own hands, not only of a male and female quagga, but also of a hybrid foal of a brood mare and quagga, which shows faint marks of stripes.

Half a century ago Captain W. Cornwallis Harris, in his magnificent folio of the "Wild Animals of Southern Africa," describes the quagga as existing in immense herds in the Cape Colony in the open and level lowlands; and, writing some seventy years since, Thomas Pringle, the well-known poet of South Africa, who was intimately acquainted with the large animals of the Cape Colony, described the quagga as then abundant in the Great Karroo. In his poem, "Afar in the Desert," he writes:

Afar in the desert I love to ride,
With the silent bushboy alone by my side;
O'er the brown Karroo, where the bleating cry
Of the springbok's fawn sounds plaintively,
And the timorous quagga's shrill whistling neigh,
Is heard by the fountain at twilight grey,
Where the zebra wantonly tosses his mane,
With wild hoof scouring the desolate plain.

And in a note he says: "The cry of the quagga (pronounced quagha or quacha) is very different from that of either the horse or ass, and I have endeavoured to express its peculiar character in the above line;" in another note to the same poem he says: "The zebra is commonly termed wilde-paard, or wild horse, by the Dutch African colonists. This animal is now scarce within the colony, but is still found in considerable herds in the northern wastes and mountains inhabited by the Bushmen."

The geographical range of the quagga appears to have been much more restricted than that of the other species. Mr. H. Bryden, in his interesting work entitled "Kloof and Karroo," which may be rightly described as an admirable account of the sports, legends, and natural history of the Cape Colony, writes as follows:

"The range of the true quagga was even more arbitrarily defined. This animal, formerly so abundant upon the far spreading karroos of the Cape Colony and the plains of the

Orange Free State, appears never to have been met with north of the Vaal river. Its actual habitat may be precisely defined as within Cape Colony, the Orange Free State, and part of Griqualand West. I do not find that it ever extended to Namaqualand and the Kalahari Desert to the west, or beyond the Kei river, the ancient eastern limit of Cape Colony to the east. In many countries, and in Southern Africa in particular, nothing is more singular than the freaks of geographical distribution of animals. A river or a desert, or a little belt of sand or timber, none of which, of themselves, could naturally oppose a complete obstacle to the animal's range, is yet found limiting thus arbitrarily the habitat of a species."

Like Burchell's zebra, the quagga was more equine than asinine in character; but it wanted the callosity on the inner side of the hind leg below the hock which is characteristic of the horse. The quagga was marked on the head and neck and front of the body with dark brown stripes on a light reddish-brown ground. These stripes gradually faded away behind the shoulder, and were absent from the hind quarters. There was a broad dark stripe down the centre of the back; the under-surface of the body, legs, and tail were nearly white. The stripes on the neck ran up into the mane, which was banded alternately with white and brown. The crest was high, the ears short, the tail fairly covered with hair, so that the animal altogether was much more closely allied in appearance to the horse than to the ass. The extermination of this animal is greatly to be regretted; it is most lamentable to know that this species, which might have become a most useful domestic quadruped, admirably fitted for the requirements of the inhabitants of the country of which it was a native, should have been shot down by the colonists merely for the sake of its hide; and it is sincerely to be hoped that its congener, Burchell's zebra, which is still in large herds to the north of the

Orange river, and which promises to be so exceedingly valuable, may be reclaimed and utilised for the service of man. Its preservation is equally desirable from a utilitarian as from a zoological point of view.

Since the foregoing account of the quagga was in type, it has been stated by Mr. Rieche, the importer of the giraffe and other South African animals, that it is possible that the true quagga has not been exterminated.

CHAPTER X.

HYBRID EQUIDÆ.

It would appear that all the different species of the genus Equus are capable of breeding together and producing hybrid offspring, some of which are perfectly sterile mules, whilst others are apparently fertile, either with one or other parent species if not *inter se*. Some of these hybrids are of great economic value, and it is deeply to be regretted that the opportunities that have presented themselves in our European zoological collections have not been utilised as they might have been, in introducing new species into the service of man, and in producing other useful hybrids beyond the common mule.

In the present chapter I propose to enumerate, as far as practicable, the various equine hybrids that have been produced, and of which any definite account has been published, commencing with those of the horse.

Horse (*E. caballus*) Hybrids.

It appears most probable, though it has not been absolutely proved, that the horse is capable of producing hybrids with every other species of the genus Equus. The hybrid between the horse and the ass is well known. When the ass is the male parent it is termed a mule; on the other hand, if the horse is the sire the produce is termed a hinny, or in some places a jennet. The consideration of the breeding and practical utilisation of these two hybrids will be fully treated of in the concluding chapters.

The horse has bred repeatedly with both the Mountain and Burchell's zebra. In the Jardin d'Acclimatation there is at the present time a hybrid between the horse and the Burchell's zebra, of bright bay colour, with black legs and distinct dorsal stripe. Some years since I described some hybrids between the horse and the female Burchell which were in the park of Sir Henry Meux at Theobalds. The sire of one was an ordinary park pony, that of the other an American trotting pony. This latter hybrid was striped on the legs, neck, and haunches. Both of them, as might be expected, showed much of the equine character and form of the male parent; and from the relative sexes of the parents they necessarily partook more of the characters of the hinny than of the mule.

Early in this century a pair of hybrids, bred between the horse and Burchell's zebra, were driven about London in the service of the Zoological Society, but I have not been able to ascertain definitely the relative sex of the two parents, but believe they were hinnys from a zebra mare. The horse has also bred with the Asiatic ass (*E. hemionus*). In a private letter Lieutenant J. L. Harrington informs me of a male Hemione breeding with an Indian pony, and producing a hybrid that, with the exception of the tail, which was asinine, looked more like a pony than anything else.

Two hybrids, between a Hemione and a mare, in the Jardin d'Acclimatation, were described by the late Mr. Jenner Weir. One of these is a very beautiful animal, possessing no shoulder stripes, and with very faint dorsal stripe.

Ass (*E. asinus*) Hybrids.

The hybrids between both sexes of the ass and the horse have been spoken of under the last heading. The ass also hybridises freely with Burchell's zebra; a hybrid of this is now in the Jardin d'Acclimatation. It is rather sparely striped, but the three shoulder stripes are well marked.

Asiatic Ass (*E. hemionus*) Hybrids.

The Asiatic ass hybridises with the horse, as has been already stated. It has also been mated with Burchell's zebra in the Jardin des Plantes, the produce being a faintly striped animal with a broad dorsal stripe, the hind quarters of which are not striped but dappled. The cross between the Asiatic ass and the mare has been already named.

Mountain Zebra (*E. zebra*) Hybrids.

Several of these were apparently recorded in the "Knowsley Menagerie," but sufficient care was not taken to distinguish between the two species, namely, the Mountain and Burchell's zebras.

Burchell's Zebra (*E. burchellii*) Hybrids.

Burchell's zebra breeds most freely with several of the other species of Equus, and there is no doubt whatever that the hybrids of this most horse-like of the asses and zebras now existing would be exceedingly valuable to man if the animals were mated as carefully as is done in breeding heavy draught mules in Poitou, and pack mules for the military service in India. The Burchell is an animal much better adapted by its structure and form to the use of man than the other wild asses, and were it properly mated and utilised would no doubt produce most valuable hybrid offspring. The hybrids of the Burchell zebra with the horse have already been mentioned; it also breeds freely with the common ass. In the Gardens of the Zoological Society at Melbourne there are some Burchell's zebras that were bred in Paris, for this most useful animal breeds freely in confinement. On September 6th, 1892, an experiment

BURCHELL'S ZEBRA AND HYBRID FOAL.

(From a photograph.)

was made by crossing the zebra with a white so-called Siamese ass, which was obviously a variety of the domesticated Equus asinus. The foal was born on October 25th, 1893, showing that the period of gestation in Burchell's zebra resembles that of the ass in being considerably over twelve months. The young one is described as a strong, vigorous animal, galloping round the enclosure when a day old and evincing considerable speed. Its colour is somewhat remarkable, not resembling that of its white sire, but being very dark with pronounced shoulder and dorsal stripes, black tips to its ears, and bars on the legs, which are well marked, especially over the joints—the zebra from which it was bred being a true Burchell, not marked on the legs like the variety known as Chapman's zebra. The foal is described as being a compact and well-made little animal, showing splendid bone. As the progeny of the Burchell zebra are likely to attract much attention, I reproduce the photograph as it was published in the *Australasian*.

In the Jardin d'Acclimatation there is another hybrid between a Burchell's zebra and a white Egyptian ass, which shows three distinct shoulder stripes, but otherwise is very faintly marked.

A hybrid between a male Burchell's zebra and the common ass was bred by the Earl of Derby and figured in the "Knowsley Menagerie." It was utilised by being driven in tandem, and the skin was afterwards deposited in the British Museum. The Hemione or Asiatic wild ass has also been bred with Burchell's zebra.

Quagga (*E. quagga*) Hybrids.

In Colonel Hamilton Smith's unpublished volume he gives a portrait, drawn by himself, of a hybrid, the foal of a quagga and a brood mare. This was faintly striped on the fore-quarters.

In the fine collection of plates known as the "Knowsley Menagerie" there are numerous illustrations of the wild Equidæ, more especially of the striped species inhabiting Africa, namely, the Equus zebra, E. burchellii, and E. quagga. All these species interbreed, not only with each other, but with the wild unstriped asses of Asia. Dr. Gray figured in the "Knowsley Menagerie" a mule bred at Knowsley between a male Tibetan wild ass, or kiang, and the female zebra. In this the legs and neck are banded. There is also a figure of a mule between a Maltese male ass and zebra, in which the head, neck, and legs are well striped, the body less so, and the hind quarters profusely spotted. Should any of my readers refer to the plate in the folio they will find that the names of those two have been transposed, as is evident on referring to the text. There are also figured a mule between Burchell's zebra and the common ass; a second between the ass and the kiang, the titles of which are also transposed on the plate; finally, we have a mule between the kiang and Burchell's zebra, and, what is very interesting, a representation of the offspring of a mule, of male ass and zebra parentage, with a bay pony mare. This strange animal may be described as iron-grey, with a short, narrow dark band on the withers, very faint indications of perpendicular stripes on the sides, distinct dark stripes on the hocks and knees, a horse-like tail, bushy from the base, and a heavy head with a grey hog mane. This creature, singular from its triple parentage, was eight hands high, and was regularly used in harness.

PART II.

MULES AND MULE BREEDING.

CHAPTER XI.

THE UTILIZATION OF MULES.

It is a remarkable circumstance that the utility of mules is a fact that requires to be demonstrated in England at the present time, although it is freely acknowledged and extensively acted upon in almost all other civilized countries employing horse labour to any great extent. In France the agricultural interest of a large portion of the west central districts mainly depends upon mule breeding, as many as 50,000 mares being annually maintained for the purpose of breeding the magnificent Poitou heavy draught mules, which command a much higher price than horses of similar standard. In Spain and Italy the employment of mules is proverbial. In America a hundred years ago mules were viewed with the same amount of prejudice that they are in England at the present time. Now, perhaps, the greater portion of the agricultural labour in that country is performed by mules, of whose advantages the acute Americans are firmly convinced by the most potent of all reasoning, that of experience, and large consignments of the best European donkeys are constantly being made to the States for the purpose of mule breeding. Our military operations when on active service cannot be carried on in foreign countries without the aid of mules, inasmuch as horses are utterly unable to endure the severe work that the animals are called upon to perform. There is no possible doubt in the minds of any persons who are acquainted with the subject, as will be fully demonstrated in the following chapters, that in endurance, capability of hard labour, economy in keep, longevity, and freedom from disease, mules far surpass horses, and it is these good qualities that have caused them to be almost universally adopted in the south of France, Spain, Italy, and, above all, by our American cousins. In the extensive wheat fields of many thousand acres which are to be found on the prairies of the United States, may be seen at one time ten or fifteen reaping machines, each one of which is drawn by a pair of mules, not a single horse or mare being visible. It may be asked then, what are the circumstances that have rendered mules hitherto so lightly appreciated in this country? The only answer to be made to this question is, that it is due to the unfounded prejudices which are based upon the most extraordinary ignorance of the merits and characters of the animal. It is difficult to conceive or overstate the want of knowledge and the false ideas that prevail regarding them, and this not only amongst persons who have little knowledge of the subject, but amongst those who are regarded as authorities upon equine subjects. Thus Mr. Robert Wallace, professor of agriculture in no less an institution than the University of Edinburgh, when writing about mules, in his valuable and practical work "Farm Live Stock," published as recently as 1893, does not appear to know whether the animals are fertile or barren, and states that:—

"The mule is generally believed to be barren, but is not so in the case of the female mule and the female hinny."

It being well-known to those who are acquainted with the subject that no satisfactorily authenticated example of a fertile female mule bred between the horse and ass has ever been known, and, as will be shown in the chapter on this subject, that in the mule breeding districts of France, where many thousand mules are produced annually, such a thing as a fertile female mule is utterly unknown, although the conditions under which the animals are kept would be favourable to such a result. Again, in a manuscript work on the Equidæ by the late Colonel Hamilton Smith, illustrated by one hundred folio drawings of the varieties of equine animals, the author states that three male mules are born to one female, a statement not worth quoting or noticing except as illustrating the prevalent ignorance regarding these animals, the proportion of births of the two sexes being about equal. But perhaps the most remarkable example of multiplied errors has recently appeared in a book entirely devoted to horses, namely "The Horse World of London," by W. J. Gordon, published by the Religious Tract Society, 1893. The writer states that:—

"There are over 200,000 donkeys in Ireland employed in agriculture, and these are of all sizes, some of the larger having a strain of horse blood in them, as is the case in Italy, where the so-called donkey is a by no means insignificant animal. Italy has more donkeys than any other European country, there being over 700,000 of them there; while France, which of late years has taken to that most difficult of pursuits, mule-breeding, has 400,000. The great mule-breeding country is, however, the United States, where there are two and a half millions of mules and donkeys taken together, it being found impossible to separate them owing to the varying proportions of horse ancestry producing an indefinite series from the genuine mule to the asinine mulatto. For the male mule is not always sterile, and the female will breed with horse or ass, or apparently any species of Equus."

It is difficult to conceive a statement respecting mules or donkeys which could be so utterly baseless as the farrago of nonsense just quoted. That the size of the Italian donkeys should depend upon their being hybrids is, of course, utterly unfounded, and that it is impossible to separate the mules and donkeys in the United States owing to the various proportions of horse ancestry, is one of the most ludicrous statements that ever was made. From such a tissue of absurdities it is most satisfactory to turn to a work which has been published by a gentleman long resident in India, and who is perfectly acquainted with the value of the mule in that country.

Mr. John L. Kipling, in his most delightful and instructive volume entitled "Beast and Man in India," informs us that the mule is of European introduction, being really a Government institution, he adds :—

"The mule, however, is bred in increasing numbers, for he is an ideal pack animal, born and made to carry the burdens of armies over difficult countries, and good at draught. Sure of foot, hard of hide, strong in constitution, frugal in diet, a first-rate weight carrier, indifferent to heat and cold, he combines the best, if the most homely, characteristics of both the noble houses from which he is descended. He fails in beauty, and his infertility is a reproach, but even ugliness has its advantages. The heavy head of the mule is a mercy to him, for both in practice and the written orders of Government, it is ordained that he is not to be bothered with bearing-reins."

From those who know the actual working of mules in other parts of the world, it is perfectly easy to get any amount of evidence as to their extraordinary value. In rough countries they far surpass any other equine animal.

A correspondent, dating from Texas some few years since, writes as follows :—

"I have just returned from a trip west with a mule train, of about 400 miles, through a country where bridges are unknown, and the roads are the best place you can find to drive—sometimes mountainous, intersected with steep-banked creeks; at others long steep rises, with draws between 2ft. or 3ft. deep in black mud, and after a rain almost impassable for miles, as the ruts cut in axle-deep, and if you leave them you have to unload and get back to where you can feel bottom. We frequently helped to pull out teams that were stuck fast, and for one mule team we pulled three horse teams, as, if properly handled, the mules will come down on their knees at a pull as many times as you ask them. We never ask but twice, and, if stuck fast, either cut loose the 'trail' or double team. In explanation of this last term, I may state that the usual way of freighting is to take four to eight mules, generally six, two abreast, the leaders small quick Spanish mules, with a span of large American mules as wheelers; the driver riding the near wheeler. Two waggons, the larger one in front, and a lighter one or 'trail' behind, are attached to the axle of the front one, so as to be easily uncoupled, and fitted with powerful 'California' breaks, which the driver controls with a line. He drives with a single rein, or 'jerk line.' Having the load divided between eight wheels, it does not cut into the sand or mud as it would on four.

"Six mules, the leaders no larger than ponies, will take 6000 to 7000 pounds anywhere, making fifteen to thirty miles a day according to the state of the roads, and I have known a team in summer driven fifty miles, with 1000 pounds a head of load, to reach water, and not appear to suffer. They do not require the feed horses do (who invariably lose flesh in the winter time), but will live on maize with very little roughness.

"With regard to drivers, you find more Irish or English than niggers; it is harder work than the latter appreciate. I should like your supply officers to have seen a train of the United States cavalry I met in January last, in six inches of snow, after a march of eighteen days through a country where they had to haul their own feed and supplies, and compare the mules with the outfits after the autumn manœuvres, as I saw them some few years ago. Mules, weight for weight, will pull more than horses, live on less, and 'come down in a tight' more times.

"Now for farm work; there is a patch of 250 acres of wheat a few miles from here, where last spring was open prairie, that was ploughed and planted with two span of mules, and looks as well as any farmer can desire. With a good sulky plough, which does not tire the driver, a span will plough two to three acres per day.

"For saddle or driving, if a man has a really good saddle mule, he is like the kings and great men of old; he would not trade for all the horses in the country. They are as pleasant to drive, and if properly handled as gentle and good-conditioned, as horses."

Another writer recounts the advantages which as beasts of burden they possess over the horse:—

"First, their working life is longer, in the ratio of about five to two, than that of a horse; secondly, they can live and thrive upon food which soon reduces a horse to a weak and helpless skeleton; thirdly, they are indifferent to heat or cold; fourthly, they never know what it is to be sick; fifthly, they can work day and night without being worn out; sixthly, they walk quicker than horses; seventhly, being light of limb and bulky of body, their weight is better disposed for moving heavy loads; eighthly, they are, when of full size, considerably stronger than a team of equal-numbered horses. I might repeat many other lesser advantages which they possess. But at a time when horse fodder of all kinds is continually rising in price, a farmer who has from ten to twenty horses to keep would soon find how much he would save in a year were he to replace them with mules; while, into the bargain, he would get twenty-five years of work out of a fine mule where it is rare for a horse to last more than from ten to twelve years."

To those interested in the subject and therefore desirous of making themselves acquainted with the facts bearing on the advantages of mule labour, the value of the mule has been long known. Mr. John Chalmers Morton, writing nearly twenty years since, speaks of the draught Poitou mules in the following terms :—

"They are hardy, willing workers, of great power, and good-tempered; they will produce and put in exercise more force per shilling of their daily cost than horses; they are less liable to injury or illness; and they are longer lived. This is 'the case' in favour of the mule as compared with the horse for farm work. It has long since been proved and known in other countries, and the powerful mules of Poitou, and mules similarly bred in America, accordingly command higher prices than are given for horses of corresponding size or for corresponding uses. It is not yet known in this country."

Col. Langhorne Wister, of Philadelphia, U.S., a great "raiser" of mules, gives the following account of the value of these animals in a private letter to a friend :

"I have made a good many inquiries about mules for work of all sorts, but especially for farm work, and find that all who have used them think them more valuable than horses. In York County, Pennsylvania, they have almost entirely taken the place of horses for farm work, and the farmers say that they can stand more work, can get along on more inferior food, and can endure infinitely more hardship than horses, and are fully as tractable. It is a very well-known fact that mules live on an average much longer than horses, and I never saw a mule, no matter how old, that could not do his ordinary work. I will not assert, what was frequently asserted before our war, that no one ever saw a dead mule, for many died during the war; but they supplanted horses entirely for draught purposes, and stood all the hardships of campaigns better. The York County mules are of large size, and usually brought in from Kentucky, the state where the best ones are bred. By large size I mean 15 to 16 hands high, and weighing about from 1000lb. to 1200lb., but often much heavier. However, I have seen teams used about the charcoal blast furnaces which would average 16 hands, weighing 1400lb. each, the tall mules being nearly 18 hands high; but of course such are not common, nor are they desirable. To sum up, I think I can say that mules live on an average five years longer, and are able to do heavy work at least seven or eight years longer than horses, they thrive on coarser food, and are more free from disease. They are very easily broken by those who understand them, but need kind treatment, as they are apt to repel force by force—*i.e.*, by kicking or striking with the fore feet."

CHAPTER XII.

NON-FERTILITY AND LACTATION IN MULES.

THE natural history of hybrids of all kinds has not received due consideration even from naturalists and scientific observers, and but little is known regarding them compared with what has been ascertained respecting their progenitors. No careful consideration of the facts relating to hybrids has been put upon record, a few scattered observations as to the fertility of some, and the absolute sterility of others, are about all that has been made known.

The extraordinary circumstance that Mr. Bartlett, superintendent of the Gardens of the Zoological Society, should have definitely ascertained that fertile hybrids can be bred between species as distinct as the bison of North America, the buffalo of India, and the wild ox of Europe, has passed almost unnoticed, although portraits of the singular triple crosses so produced have been published in the Proceedings of the Society. But no further experiments have been made with the view of introducing either of these crosses into our breeds of domestic cattle, with the possibility of improving the characters of the latter, and at present the only advantage that has been gained by Mr. Bartlett's interesting experiments has been to ascertain the fact that three very distinct species of the Bovidæ, inhabiting different parts of the world, can be bred together in almost any manner so as to produce fertile compound hybrids.

Regarding the facts that more immediately concern us, the character of the hybrids between the horse and the ass, much more has been ascertained, although little scientific observation has been brought to bear upon the question. The relative influence of the male and female parent in these cases is now well known, and the distinction between the mule (the offspring of the ass and the mare) and the hinny (the result of the union of the horse and the she ass) is well ascertained. Both offspring depend for their size on that of the female parent. As far as is known from accurate observation, male and female mules and hinnys are absolutely sterile, although certain accounts of fertile female mules have occasionally appeared in print.

Captain Hayes, a very practical authority, writing on this subject states:

> "Neither the mule (the produce of the jackass and mare) nor the hinny or jennet (the cross between the horse and the she ass) is fertile, either among themselves, or with other members of the horse family. Those animals which have been mistaken by superficial observers as fertile mules, have been, I venture to say, in most cases the offspring of mares that have previously bred to donkeys, and have endowed their young with some of the characteristics of their former asinine lovers. Both the mule and the jennet respectively 'take after' their dam in size, and their sire in appearance and disposition."

Those persons who have paid the greatest amount of attention to mule production and mule industry know of no instance of a female mule producing young, and M. Ayrault, in his valuable treatise "De l'Industrie Mulassière," the standard work on mule breeding in France, says that in Poitou, where 50,000 mares are annually employed in

SUPPOSITITIOUS MULE AT JARDIN D'ACCLIMATATION.

breeding mules, such a thing as a fertile mule is unknown, although these young mules are placed in the most favourable conditions for being mated, as they are constantly in the pastures and on the marshes with the young horse colts. M. Ayrault's exact words are as follows :—

"Nous ne rechercherons pas ce que cette opinion peut avoir de fondé, mais ce que nous tenons à constater c'est que jamais en Poitou on n'a entendu parler de la gestation de la mule, bien que là, à part la temperature, elle se trouve dans les meilleures conditions pour être fecondée, puisqu'elle est constamment en contact, dans les pasturages, avec des poulains (horse colts), qui souvent les saillissent." (p. 152.)

To this it may be replied that there is a well-known instance in the Acclimatisation Gardens in Paris, where a mule has produced foals when mated both with the horse and the ass. As this is supposed to be the most authentic case on record, it has been thought desirable to reproduce from a photograph an exact representation of this supposed fertile female mule, which has been most carefully drawn by Mr. Frohawk. It is doubtful whether the animal is a mule. There is but little mule character about her beyond the slight increase in the size of the ears. The particulars of her parentage are utterly unknown, and she was merely alleged to be a mule by the Algerian natives who sold her to the authorities in the gardens. It is not at all improbable that her female parent had bred a mule in the first instance, and, as in the well-known cases of mares which have been mated with quaggas and zebras, her subsequent progeny, when mated with a horse, shows some trace of the first union. The late M. Ayrault, and most persons who are really cognisant of the matter, regard this animal not as a mule, but as an ordinary mare. She has foaled both to the ass and the horse. Her foals bred from the ass appear to be ordinary mules, and are sterile, whereas if she were a mule they should be three-fourths asinine and only one-fourth equine, which is not the case. Her progeny by the horse are horses which have proved fertile. It would appear most probable that this is not a case of a fertile mule breeding; but, that the animal is really an ordinary mare, whose female parent was influenced by a first alliance, as is so often the case in dogs and other animals.

There is no doubt that the majority of the accounts of supposed fertile mules owe their origin to the fact that abnormal lactation not unfrequently occurs in them, when milk is secreted in great abundance, and they may be seen suckling the foals of other animals. This singular phenomenon is not confined to mules, but is well known to occur in many other species.

The maternal instinct is one of the most powerful, and there are numerous examples of its being so strongly excited in females (other than the mother) in favour of the young of animals of the same, and even of different species, as to determine the abundant secretion of milk. Domestic animals, such as cats and dogs, have been known to suckle young of other species, even when they had no progeny of their own; and corresponding instances among women who have fostered orphan children are on record in the physiological journals. Nay, more than this, a case is related by Humboldt of a man who became the wet nurse to an infant child. "In the village of Arenas there lived a labourer, Francisco Lozano, who had suckled a child. Its mother happening to be sick, he took it, and, in order to quiet it, pressed it to his breast, when the stimulus imparted by the sucking of the child caused a flow of milk. The man was examined by M. Bonpland, who found the breasts wrinkled, like those of women who have nursed.

He was not an Indian, but a white, descended from European parents." Other authors have given examples of the same nature.

The late Mr. Francis Francis described in the *Field* for Oct. 27, 1860, a maiden bitch at the Vine Kennels, that brought up two litters of puppies in succession, and he saw the last when they were about six weeks old. This communication called forth several letters giving other examples of similar facts. Mr. Sprent, of Reading, writing, stated that he had a terrier bitch that never had puppies, but she took a kitten from its mother, and had a good secretion of milk with which she nourished it, and numerous examples of similar facts are on record. The most important, as having a direct bearing on the subject, was recorded in a letter printed by one of the authors of this work, Mr. W. B. Tegetmeier, in the *Field* of April 17, 1880, in which he says:—

"The case, however, which I am about to put upon record is, I think, unprecedented, inasmuch as it is that of a sterile hybrid animal suckling another. The facts are as follows: An aged brown female mule that formerly, when in the possession of Messrs. Flower, of Stratford-on-Avon, had taken prizes at the large shows as a heavy draught mule, passed into the stables of Mr. Cole, of Church-street, Chelsea, who is well known as one who has employed mule labour with great advantage for many years. I accompanied Mr. C. L. Sutherland to the stables of Mr. Cole, where we saw the mule in question, and a young male donkey nearly a year old. This donkey foal had been bought, and allowed to run about the stable yard. It had been noticed to follow the mule, and at night to go into her stall at the further end of the stable, where he was observed sucking the mule, whose udder, on examination, was found fully charged with milk. Thinking the proceeding rather 'unnatural,' Mr. Cole had the donkey removed, and the mule milked by hand; but this was not done to a sufficient extent, and, in consequence, milk abscess occurred, which opened, the udder having previously swollen to a very large size.

"This case is interesting, inasmuch as it proves that the secretion of milk can take place in a hybrid animal which is naturally sterile, and that it has no necessary connection with the maternal relations."

This example is not by any means a solitary one. A communication from the late Mr. J. B. Evans, of the Cape Colony, appeared in the *Live Stock Journal* of June 23, 1893, in which he spoke of a mare mule having to be milked, as each year she had adopted a foal, driving the mother away, and secreting milk in abundance for the support of the foal that she had fostered. Accounts not unfrequently appear in the American and other papers, of mules which are seen suckling young, and the conclusion is at once arrived at that these young are the offspring of the animals that are supporting them, but it may be regarded as perfectly certain that they are merely adopted foals, which by their endeavours to suck female mules have developed in the latter abnormal lactation.

Brown Poitou Mule (16 hands).

CHAPTER XIII.

THE POITOU MULE.

The marked distinction between the different types of mules that are used for heavy draught, and the lighter varieties that are employed for riding and the army service, renders it desirable that they should be considered in different chapters. As an animal for agricultural use the Poitou mule far exceeds in value any other breed, and it would be desirable to consider it, and its progenitor the Poitou ass, in the first instance.

The old province of Poitou in the west of France has, agriculturally speaking, for some centuries given itself up in a great measure to the breeding of mules, chiefly for the market. Extreme care has been taken in the breeding of the asses, the sires of these mules; and the Poitou mule fairs—especially those held in the winter—are attended by foreigners from all parts of the world for the purpose of buying the great local production, the mules. It is not an uncommon sight to see at a fair as many as 1000 mules, from one to four years old, offered for sale.

There may be said to be two types of mules bred in the Poitou, the light and the heavy, but the latter largely predominate. The breeders find that heavy mules are more in request, and bring more money as beasts of draught than the lighter animals, and consequently they endeavour to produce mules as *gros* as possible. The finest and largest cart mares are selected for the purpose; indeed, the best mares are always put to the ass (or *baudet*, as he is termed) in preference to the horse. A mare if she is capable of breeding a mule is considered more valuable than one which will only breed to a horse. All mares are not what is termed *intérieurement mulassières*, and in that case they are used to breed horse colts from.

The peculiarities of the Poitou mule as distinguished from the Spanish and other mules are as follows:

The Poitou mule is eminently qualified for service as a beast of heavy draught, and as such is capable of taking the place of any ordinary farm horse. The head and ears are large and decidedly coarse, according to our notions; but the Poitou breeders maintain that they cannot get the necessary weight of barrel without a correspondingly large head and ears. The neck is short, and the animal often carries a good crest. The chest is broad, the shoulders rather upright and muscular. The mule is often a little longer in the back than is desirable in a draught animal, and is apt to stand over too much ground. The barrel is capacious and well let down, though sometimes the sides are apt to be a little flat. The quarters and thighs, while strong and muscular, present on the whole a narrower and lighter appearance than those of a draught horse, and it is in these points particularly that Poitou mules require improvement. The hocks are large, and, while a large proportion of mules are cow-hocked, this conformation does not render them more than ordinarily liable to throw out bony growths, or to suffer from strains of tendons or ligaments. While on the subject of bony growths, it may be as well to correct a very prevalent idea that mules are not as subject to them as horses. It is quite true that they are not to the same extent; but specimens of spavins, ring-bones, side-bones, and splints, as well as curbs, may occasionally be seen in mules. The peculiarity is that, although these exostoses are in many cases well developed, the animals, owing to a singular want of sensitiveness, rarely go lame with them.

We now come to the distinguishing characteristic of the Poitou mule, viz., limbs and feet. The legs are short and stout, with plenty of bone, and the pasterns short, as becomes a draught animal, and there is sometimes a good deal of hair about the legs. By the limbs being stout it should not be understood that they are round and gummy; on the contrary, they are flat and hard, whilst the feet are larger and more expanded than those of any other breed of mules. The heels are in many cases somewhat contracted; but in breeding this can be obviated to a great extent by selecting a good open-footed jack as a sire.

It is in the matter of feet and limbs that the Poitou differs essentially from the Spanish mule. It is well-known to mule-breeders that, in crossing jacks and mares, the resulting mule will take after the ass, its sire, in all its extremities—that is, in ears, legs, feet, and tail. Thus in Spain, where the asses are much finer in their limbs than they are in Poitou, it is no uncommon sight to see mules which may be aptly described as animals having a horse's body on a donkey's legs and feet. The result is, that animals of this conformation are utterly incapable of steadying a heavy load on a bad road when placed in the shafts, and, being swayed about by their load on account of their barrels being too large for their limbs, their legs and feet "give out" as the Americans term it. The value of a good-sized foot for travelling over deep, heavy land must also be taken into consideration.

The Poitou breeders, having made this discovery, have for some centuries devoted themselves to rearing a breed of asses as mule-getters with as large limbs and feet as possible, and the consequence is that the Poitou mules are much more symmetrical in form and appearance generally, and more capable of moving a heavy load, than the Spanish. The Spanish mule is better fitted for light trotting work than the Poitou, but it is the latter animal which is pre-eminently suited for introduction into this country for agricultural purposes as an auxiliary of, and substitute for, the horse.

Poitou mules are of all colours—bay, brown, black, grey, white, and sometimes chestnut and skewbald; but about four-fifths of them take after the *baudet* their sire in colour, and he is always black, or dark brown. The height of the draught mules ranges from 15 to 16 hands, rarely more. Spanish mules sometimes reach 17 hands, but there is generally too much daylight under these very tall animals. The females always realise higher prices than the males, chiefly on account of less risk being supposed to attach to them during sea voyages. The price of a good draught mule of three or four years of age ranges from £40 to £60, sometimes reaching as high as £80; whilst a draught horse or mare of corresponding quality and capabilities can be purchased for from £30 to £40.

The engraving at the head of this chapter represents a brown mule which obtained a prize at the Grand Concours Mulassier, held at Niort, in the Deux Sévres. She was the property of M. Auguste Disleau, of Sainte-Ouenne, and stood just 16 hands at four years old.

The second engraving is a copy of a photograph taken at the Bath and West of England Show, Croydon, of Brunette, an inported Poitou mule 16.1, belonging to

Poitou Mule Brunette. (Height 16.1.)

Mr. C. L. Sutherland. She may be regarded as a typical specimen of a first-class draught mule in working condition, and won the following prizes:

First prize Bath and West of England, Croydon, 1875.
First prize Royal Agricultural Society, Taunton, 1875.
Second prize Crystal Palace, 1875.
Second prize Alexandra Palace, 1875.
First prize Dairy Show, London, 1877.
Third prize Royal Agricultural Society, Kilburn, 1879.
First prize Alexandra Park, 1881.

When speaking of the large limbs and feet of mules of the Poitou race, it must not be understood to signify that they are as large as those of a cart horse of corresponding height, but as speaking comparatively, and looking at the limbs and feet of the generality of mules. A mule can never be a horse, and it is only by the careful selection of asses as sires, with points approximating as nearly as possible to those we look for in a horse, that we can expect to breed symmetrical mules. It is quite true that almost any mare, coupled with any ass, will produce an average mule; but if we wish to breed first-rate animals we must take special care in selecting sires and dams. The grey French cart mares from which the Poitou mules are bred are very middling animals when compared with our Shire and Clydesdale breeds; yet they give good produce when coupled with the Poitou ass. If good Poitou asses were selected and used on our English cart mares, there can be no doubt we should produce mules far surpassing any yet bred in Poitou. The Americans breed mules from their best cart mares, and find it pays them better to do so than to breed horse colts. They mostly use, however, for the purpose the tall Spanish ass, originally brought from Catalonia, leggy, light in the barrel, and generally with small limbs and feet. The Maltese ass, which is also occasionally used, may be described as an "improved Spanish," and is certainly a better animal for the purpose than the normal Spanish ass. If they were to use the Poitou instead of the Spanish or Maltese ass, they would obtain very different results.

Americans have expressed astonishment at the size of the limbs and feet of the mules in Poitou, and admitted that this was a deficiency in the limbs and feet of the American mule which required correcting.

The late M. Eugène Ayrault, the most intelligent veterinary surgeon at Niort, author of "De l'Industrie Mulassière," computed that in Poitou 50,000 mares are employed in the mule-producing business, of which number 38,000 are mated with the ass, and the remainder with the horse; in fact, mules and mule-breeding are about the only things talked of in agricultural Poitou. Every farmer, every peasant, every petty proprietor breeds a mule or two, which he knows he is quite certain of selling at a remunerative price at any of the numerous fairs, where the relative value of mules and horses may be pretty nearly arrived at from the fact that, while a charge of twenty centimes is, as a general rule, made for the right to take a horse for sale on to the *Place*, or wherever the fair may be held, a charge of thirty centimes is made for a mule.

A great many of the so-called Spanish mules seen in the neighbourhood of the Pyrenees and in the North of Spain are in reality Poitou mules, the Spaniards always attending the Poitou fairs in large numbers for the purpose of buying the mules having the most *distinction*. The mule merchants from the South of France also buy thousands of mules, which they take with them to Marseilles, Montpellier, Toulouse, &c., where the animals bring high prices.

The Americans also attend the fairs, and buy many mules, which they export from Nantes and St. Nazaire. It may be asked, why don't the Americans buy the asses too? For the reason that the male asses are not brought to the fairs. They are a great deal too valuable to be exposed for public sale, and are disposed of privately, and then only with the greatest possible form and ceremony.

The principal mule fairs are held in the winter (in January and February), the mules having been for some two or three months previously released from work and got as fat as possible for sale. In very many cases, however, the country has been previously scoured by the Spaniards and *marchands du Midi*, who readily buy all the good animals they can lay their hands on. The best mules are generally to be procured at these winter fairs. In the summer fairs, which are held only very occasionally, it is as hard to find one really good mule as it is to find a hundred in the winter; but the transport is of course much easier in summer than in winter.

As a general rule, the mules in Poitou are by no means well "done;" on the contrary, they are poorly fed, and hardly worked. They are broken at two years old and worked till they are three or four, when they are fed up and sold. If they were fed in proportion to the work got out of them, or if they were not quite so hardly worked, they would grow into much finer animals than they do. It is amusing to see the manner in which these mules are broken. At two years old their education commences, and it is no uncommon sight to see eight young mules harnessed to a cart, one in front of the other (with an old horse in the shafts, termed the *limonier*), belonging to eight different proprietors, each one carefully leading his own animal, alternately caressing and swearing at him. The mule being naturally nervous and timid it is necessary to exercise great patience and kindness in breaking him. It is this nervousness which is so often mistaken for vice by the ignorant, and which has given the mule a bad name with those who, not having studied his nature, have often turned a really tractable though nervous animal into a dangerous vicious brute by beating and ill-using him. Patience and kindness, combined with firmness and a knowledge of the animal's nature, will almost always succeed where brutality has been exercised in vain, and long and careful observation proves that the Poitou ass, coupled with English cart mares, would give us mules which, with our system of feeding and management, would furnish the farmer, the brewer, the coal merchant, the miller, the timber merchant, the owner of barges (mules are far better than horses for towing), &c., with the most economical form of horse labour possible.

It is well known to agriculturists that, however comparatively light and easy in the draught a reaping machine may be, no one pair of horses can go on working it all day without change. A horse sickens of always having his shoulders home in the collar, and prefers work of more give-and-take character. Not so the mule. He will go plodding on all day and every day, unceasingly, and heavy draught mules, with the necessary weight, are very valuable for this purpose—a point worthy of consideration and trial by enterprising agriculturists and machine makers.

The saving that would be effected if mules were more generally used in our army transport service instead of horses it is hardly necessary to point out. The use of mules for transport and for ambulance waggons could not

English Mule. (16 hands. From Cart Mare and Andalusian Jack.)

fail to be of advantage to the State. It is true that our men would have to be instructed in the treatment and management of them at first; but we have plenty of men in the service who have been used to mules at Gibraltar, the Cape, India, &c., who would soon impart their experience to their comrades.

The large, heavy sixteen hands draught mule bred in Poitou, such as is suitable for agricultural purposes and heavy road work, is not the only kind of mule to be found at the fairs. Mules are offered for sale of all sizes, from thirteen to sixteen hands, suitable for all purposes, whether for carriage work (for which purpose the Spaniards buy the best bred and finest in the limb), or for heavy farm or road work, or for burden, or for army purposes.

A draught mule, bred by Mr. A. J. Scott, of Rotherfield Park, Alton, Hants, from an English cart mare by an Andalusian jack, is shown in the engraving. She is now in the possession of Mr. Sutherland. She is a powerful beast, a quiet good worker, and exceptionally well formed in the hind quarters.

The mule is little appreciated in England, because it is rarely seen here in perfection. By the term mule is generally signified an under-sized, chance-begotten animal, of perhaps thirteen hands or so; and it is not uncommonly supposed that this is the kind of animal it is at the present time sought to introduce into England to do the work of a draught horse! Any kind of mule can be bred to order, by a judicious selection of sire and dam, whether it is to be a light trotting mule, fit to run between the shafts of a sulky, or a heavy draught mule, that at a dead pull will beat any horse that ever was foaled. You can get to the bottom of a draught horse by putting a weight behind him that he cannot possibly start. Such a horse, in nine cases out of ten, will never try his utmost again; and exactly the same result occurs in riding a horse to a standstill with hounds. Not so, however, with the mule. You may load him as much as you like, whether on his own back or on wheels, and, if properly managed, he will always go down on his very knees and do his utmost; and, if unable to move his load to-day, will try just as hard to-morrow.

Unnecessary brutality is often brought into play in breaking and using mules. There is no necessity for anything of the kind. There is a certain amount of firmness, decision, and patience required, but no brutality, which only engenders vice, which will show itself in an old mule that has been habitually ill-treated. The animal's nature should be studied. He is affectionate and quick in perception, but nervous and afraid of strangers. This is the first thing to recollect in dealing with mules. Make friends with him, speak to him kindly whenever you approach him, feed him a little every day, and in a week you may do what you like with him. Mules, so nervous from having been ill-treated that it is not safe for anyone ignorant of their nature to go near them, by kind and at the same time firm treatment, as a rule, become perfectly quiet and tractable.

CHAPTER XIV.

THE POITOU ASS AS A SIRE OF MULES.

The breeding of mules is one of the most curious and interesting, as well as most lucrative, local occupations of France. It is so purely local that perhaps not more than one out of every six Frenchmen is even aware of its existence, or at all events of the magnitude of the transactions connected with it. Poitou is a part of France little frequented by tourists, French or English, and consequently little is known of its productions. During the time the winter fairs are being held the country is overrun by dealers—French, Spanish, and Italian, &c., who attend the fairs, and buy strings of mules to take home to their respective countries.

The Poitou ass is supposed to have been originally of Spanish extraction. He differs, however, very considerably in outward appearance from his Spanish progenitor—a difference brought about chiefly by selection and careful breeding. His head and ears are enormous, and the larger they are the more valuable is the animal considered by the breeders. His ears are often so enormous that he is unable to carry them in an upright position. They are then carried horizontally, like those of an oar-lopped rabbit, giving the animal a most extraordinary appearance when viewed from the front. The interior of the ears, from the tip to the point of insertion in the head, is well furnished with silky ringlets, termed *cadenettes*—a great sign of purity of breed. His lips are curiously pendulous, the lower lip especially. He often carries a good long mane and forelock. His neck, while neither long nor short, is strong, thick, and broad. As in the asinine race generally, there is a want of withers, and the back is very straight. His shoulders are tolerable, inclined to be upright rather than the reverse; his chest is broad, and his limbs are simply enormous. It is in the matter of limbs and feet that the Poitou ass differs essentially from other breeds, and it is these points that the mule breeder has chiefly to regard in selecting a *baudet*. His forearm, while large, invariably exhibits a want of muscular development, owing to these animals—the males at least—never being worked or even exercised. His knees are very large, and he should "tape" well below the knee. Many Poitou jacks measure 9 inches below the knee, after allowing for hair, of which there is abundance. Eight and a half inches, however, may be considered as good measurement, the bone being usually good and flat. His pasterns are short, and his feet larger and much less contracted at the heels than those of other breeds of asses; while the feet and posterior part of the fetlock, immediately below what are known to veterinarians as the sesamoid bones, should be well covered with abundance of long silky hair, when the animal is said to be *bien talonné*. His tail is short, and usually furnished with long hair at the extremity only. His quarters are generally thin and spare, and this is a point in which he requires improvement. His body is long, and, if his ribs are not as well sprung as those of a horse, he mostly girths well. Contrary to our ideas on the subject of cart-horse breeding at least, the longer the body the better mules he is thought to produce. The bray of the Poitou ass is peculiarly loud and sonorous, totally different from that of the Spanish or Maltese breeds. The height of the Poitou ass varies from 13½ to 15 hands, and the colour is always black or dark brown. Greys are sometimes produced, but they are always rejected for breeding purposes. Height in the ass is not nearly so much looked for by breeders as the other properties of head, ears, limbs, feet, and barrel. Height is got from the dam, the mule-producing mares generally standing from 15 to 16 hands, and sometimes higher.

Perhaps the most extraordinary part of the Poitou jack is the coat, with which he is blessed or cursed, as the case may be. From the day he is born to the day of his death no brush or comb is ever allowed to be used on him; and, as from the unnatural condition in which he is kept, he is prevented in a great measure from shedding his coat, the functions of the skin become suspended, and the animal gradually assumes year after year an accumulation of coats all matted together with stable filth, till at length they almost trail on the ground! When he has assumed this extraordinary and bear-like appearance, he is pointed to with no little pride by his owner, and is termed *Bourailloux* or sometimes *Guenilloux*. Such is ignorance and prejudice! Suffice it to say that this state of things almost invariably produces cutaneous affections of the worst description. This power to retain in a great measure coat after coat is not possessed (happily for them) in an equal degree by all Poitou asses. It appears to be a peculiarity of a small minority only, and is considered by M. Ayrault to be of no practical utility whatever. In fact, he looks upon the variety *Bourailloux* as quite an inferior

SMOOTH-COATED "POITOU" JACK.

one, and liable to be light-limbed and small-footed. Still the object of each proprietor is always to have asses with as much unkempt coat as possible, be they *Bourailloux* or otherwise.

The illustration on the opposite page represents a short-coated Poitou jack, or one that, in the eyes of the breeder, has had the misfortune to lose his coat at an early age. As such he is much better adapted for breeding mules for hot climates.

The point in the Poitou ass to which exception will be taken is the great size of the head and the length of the ears. In a horse we most of us look at his head first; and a small, blood-like head, well set on, makes up for a multitude of sins. Now, the Poitou jackass is kept mainly for breeding heavy draught mules; and it has long been an axiom among the breeders that these mules cannot be produced with the necessary size and weight without correspondingly large heads and ears, and that these can only be communicated through the medium of jackasses already blessed with an excess of these appendages. Consequently, it is the aim and object of the breeders to produce asses with the largest heads and ears possible.

Like the Arabs with their mares, the Poitou breeders manifest considerable reluctance at parting with their asses, which is not to be wondered at considering the large sums of money which this mule breeding and selling brings them in. There is also not a little difficulty in rearing and bringing the asses to maturity, owing to the very false principles on which the breeders and their forefathers have proceeded for centuries.

The breeding of the asses is quite a distinct branch of the industry from that of the mules, and is almost entirely confined to the neighbourhood of Melle and Chef-Boutonne, although in most of the *ateliers* (the name given to the establishments where the jackasses are kept) one or two female asses for breeding purposes are generally to be found.

The female ass is kept entirely for breeding asses. What is known in England, and commonly in Ireland, as a mute, jennet, or hinny (the sire in this case being a horse or pony, and the dam a donkey), is never, or very rarely, seen in Poitou. An animal bred in this way is termed *bardot*, and is considered of little value in comparison with a mule proper, bred from a male ass and female horse.

In Poitou the same points are looked for in the female as in the male ass, viz., girth, large head and ears, plenty of bone in the legs, open feet and rough coats. The females are not so high as the males as a rule, and may be said to vary from 13 to 14 hands. It is scarcely necessary to mention that, looking at the relative value of male and female asses, it is the great anxiety of the breeder that his female asses shall produce male offspring. With this view, the wretched jennies are kept in as low condition as possible, under the idea that such a condition favours the production of male offspring. Indeed, the poor wretches are mostly mere skin and bone, and are supplied with nothing but hay and straw in just sufficient quantity to keep them from absolute starvation. This is another of the Poitou practices which requires sweeping away. The great wonder is that, looking at the extraordinary prejudices which prevail in Poitou, detrimental alike to animal health and animal life, the breeders yet contrive to bring into the market such fine mules as are to be seen in hundreds at the winter fairs. I use the word "mules" advisedly, as be it recollected that the breeding of the asses is only to be regarded as a means to an end, which

Yearling Poitou Jack.

end is the production of mules for the market. The breeders will sell their mules readily enough, but think twice before selling their asses.

The engraving of a yearling Poitou jack is from a drawing, admirably executed by the late Mr. T. W. Wood, of a young Poitevin *baudet.* He stood 46in., and ultimately reached 14 hands. He was very gentle and tractable in temper.

Long before the expected time of parturition the farmer or his son always sleeps in the stable, so as not to be taken by surprise, and the greatest excitement prevails throughout the whole establishment. If the young animal proves to be a female the excitement subsides quickly enough, but if a male (technically termed *fedon*) makes his appearance, great rejoicing is the consequence, and for a whole month the proprietor scarcely leaves his treasure either by night or day. But here again prejudice and ignorance step in. The young animal is deprived of the first milk, or what is known as the *colostrum,* of its mother. The peculiarly laxative effect of this milk has been well ascertained, but the Poitou peasant chooses to designate it as poison; and the young animals are not allowed to partake of what has been specially designed by Nature for their well-being, and the consequence is that in the first month of their existence the whole system becomes thoroughly deranged, and a great many of them are lost. After the first month is over the critical time has passed, and there is then little difficulty attending their rearing. Weaning takes place at eight or nine months. Those that the breeder does not require are readily bought by the dealers who scour the country, and who resell them to the keepers of *ateliers* in various parts of the province. In the case of one breeder selling a young *baudet,* or male ass, to another, or in the case of change of ownership of an adult *baudet*, great form and ceremony are attached to the transaction.

The female asses are sometimes, though rarely, employed in the agricultural labours of the farm. As a general rule, they are kept solely for breeding purposes, as there is an idea in Poitou that pregnant animals should not be worked. Possibly the breeders have at some time discovered to their cost that starvation, pregnancy, and hard work taken in combination are disposed to have a deleterious effect upon the animal system generally.

The number of *ateliers*, or mule-breeding establishments, in Poitou amounts to nearly two hundred, the majority being in the department of the Deux-Sevres. These establishments are tenanted by many hundred male asses, female asses, and entire draught horses, the latter called *etalons mulassiers*, and used for keeping up the mule-breeding race of horses and mares. The mares from which the mules are bred belong to farmers and peasants in the neighbourhood, and are brought to the *ateliers* when necessary. Each *atelier* contains from three to eight male asses, one or two females, and two entire draught horses, one of which is technically called a *boute-en-train*.

The following are the measurements of a Poitou jenny ass brought over to England for breeding purposes by Mr. C. L. Sutherland:

Height, 14 hands ½in.	Below hock, 10in.
Forearm, 19in.	Greatest girth, 77in.
Knee, 13in.	Girth behind shoulder, 66in.
Below knee, 8½in.	Length of head, 28in.
Foot, 18in.	Length of ear, 15in.
Hock, 16in.	Ears, tip to tip across, 34in.

The engraving of a Poitou female is from the photograph of a jenny of fourteen hands, which obtained the first prize

POITOU JENNY. (14 hands.)

and silver medal at the Grand Concours Mulassier held at Niort.

The following are the measurements of a Poitou jack imported by the late Mr. Ed. Pease for the purpose of breeding draught mules; they may be regarded as those of a fair specimen of the Poitou ass:

Height, 14 hands 1in.
Forearm, $19\frac{1}{2}$in.
Knee, 15in.
Below knee, $8\frac{1}{2}$in.
Hock, $17\frac{1}{2}$in.
Below hock, 12in.

Greatest girth, 77in.
Girth behind shoulder, 66in.
Length of head, 25in.
Length of ear, 15in.
Ears, tip to tip across, 32in.

The kind of mare from which the large draught mules are bred is known as *la jument Poitevine mulassière.* From official statistics, published some years ago, it appears that there were at that time 50,000 mares employed for mule breeding in Poitou, of which number 38,000 were devoted to producing mules, and the remaining 12,000 used for keeping up the breed of horses called *race chevaline mulassière.* The above number is probably exceeded in the present day, in consequence of the lucrative nature of the business, the mules costing but little to breed and rear, and realising high prices when brought to the fairs to be sold.

It may be as well to mention that from time to time the French Government has tried to discourage the breeding of mules; but, in spite of all efforts to the contrary, the business has increased year by year; mules have become dearer and dearer, they have been more and more sought after by foreigners from almost all parts of the world, and more money has consequently been brought into the country. Years ago the sum annually realised by the sale of Poitou mules was estimated by M. Ayrault at something like eleven millions of francs, equal to 440,000*l.*, and the average price, as then taken, was decidedly low in comparison with the prices brought by the mules in the present day. Altogether, mule breeding may be considered one of the most remunerative industries of France, although little is known of it outside its own immediate district.

The following engraving represents a prize Poitou cartmare and mule foal, the latter between two and three months old. The mare may be taken as a fair specimen of the race from which the mules are mostly bred. About sixteen hands high, she possesses the chief qualifications looked for by the French breeders, who make a great point of plenty of hair about the pasterns and feet—a matter of quite minor importance.

For many years the Poitou race of horses and mares was alone supposed to possess the qualifications for producing fine mules. The mares were said to be specially adapted for breeding with the ass; in other words, they were alone considered to be, according to Jacques Bujault, *intérieurement mulassières*. Naturally it was to the interest of the Poitou breeders that this fallacy should be maintained as long as possible, and for many years it was kept up most successfully. In process of time, however, means of communication improved, and mares were introduced from Normandy and Brittany. Of late years some of the more spirited breeders have imported draught entire horses from England and Belgium, with the view of improving the breed of horses. At first the old breeders were of opinion that by these means the mule-breeding business would be ruined; but experience has proved, as it naturally would prove, that finer mules than ever are produced, owing to the judicious steps taken in the matter. The Americans have perhaps more than any other

Poitou Draught Mare and Mule Foal.

nation disproved the idea of the Poitou mare being solely adapted for breeding mules; witness the magnificent animals to be seen in nearly all the States, but notably in Kentucky, Missouri, and also in New Orleans.

Each farm in Poitou includes from three to eight mule-breeding mares, according to the means of the proprietor. These animals are very rarely used in the labours of the farm, which are performed by oxen and young mules. The mares are generally kept solely for breeding purposes —for breeding mules, if possible; failing that, for keeping up the breed of horses. A mare is commonly a mother before she is three years old. If the two-year-old filly happens to prove in foal, she is insufficiently nourished on straw, chaff, and a little hay perhaps, under the idea that low condition is desirable during the period of gestation, and that starvation conduces to successful parturition; utterly disregarding the patent and common sense fact that at such times the mare requires extra nutriment for the support of herself and for the proper development of the fœtus. Then, too, at the birth of the foal, be it mule or horse, the young animal, is deprived of the first milk or *colostrum* of its mother—a proceeding which in very many cases leads to the speedy death of the foal, in consequence of a peculiar disease attacking the kidneys, and terminating fatally, unless skilled professional assistance is at once obtained. Such is prejudice and custom! Happily for the peasant's own benefit, these crude notions have been disproved owing to the exertions of MM. Ayrault, Levrier, and other skilful and influential veterinarians in Poitou.

The young mule figured would grow into a fine animal for draught purposes, and would probably make sixteen hands or more. Young mules may be seen that promise better, so far as bone in the leg, large feet, and weight of barrel are concerned, but the present animal, like her dam, may be taken as a fair specimen of the race. Her exact future it would be difficult to foretell; but one thing is quite certain, and that is that, like all her fellows, she will not end her days in Poitou. She will probably be sold so soon as she is weaned to some peasant in a part of Poitou where mules are not much bred, but only reared as yearlings; possibly again, at two years old, to another peasant, in a district where only two-year-old mules are reared; and certainly again, at three, four, or five, she will be finally sold to one of the numerous mule merchants from the South of France, Spain, or Italy. The Spaniards buy the light-trotting mules with style and good action to run in their carriages; and the *marchands du Midi*, buy the heavy draught mules. An experienced breeder on the birth of a mule foal can, and often does, foretell its future destination to an absolute certainty, according to its make and shape, *i.e.*, whether it will go to Spain or Le Midi.

CHAPTER XV.

THE AMERICAN MULE.

THE history of the mule in the United States is one which could advantageously be studied by the inhabitants of this country. At the latter part of the last century the mule was as little appreciated in America as it is in England at the present time. But little trouble, forethought or intelligence was brought to bear on the breeding of this useful animal, and the result was that but poor specimens were produced. But labour at that time was of such high value in the sparsely populated country, that the advantages of the mule as a beast of draught as well as burden were soon perceived, and great care was taken in breeding the mule from a better class of jack, and from superior well-bred mares.

It is remarkable that one of the first persons to advocate the employment of mules in the United States was General Washington. By the kindness of Sir Walter Gilbey, we are enabled to reproduce an advertisement printed by Washington in a Philadelphia paper for 1786, before his election to the presidency. It appears that the King of Spain presented him with a large Spanish jack, which by Washington was named Royal Gift. His advertisement reads as follows:

"ROYAL GIFT.—A Jack Ass of the first race in the kingdom of Spain will cover mares and jennies (the asses) at Mount Vernon the ensuing spring. The first for ten, the latter for fifteen pounds the season. Royal Gift is four years old, is between 14 1-half and 15 hands high, and will grow, it is said, till he is twenty or twenty-five years of age. He is very bony and stout made, of a dark colour, with light belly and legs. The advantages, which are many, to be derived from the propagation of asses from this animal (the first of the kind that ever was in North America), and the usefulness of mules bred from a Jack of his size, either for the road or team, are well known to those who are acquainted with this mongrel race. For the information of those who are not, it may be enough to add, that their great strength, longevity, hardiness, and cheap support, give them a preference of horses that is scarcely to be imagined. As the Jack is young, and the General has many mares of his own to put to him, a limited number only will be received from others, and these entered in the order they are offered. Letters directed to the subscriber, by the post or otherwise, under cover to the General, will be entered on the day they are received, till the number is completed, of which the writers shall be informed, to prevent trouble or expense to them.

JOHN FAIRFAX, Overseer."

"February 23, 1786."

This advertisement shows that the General was fully aware of the advantages derived from the use of mules and of the character of the jack from which they should be bred. Washington continued the use of mules during his life, and amongst the stock mentioned in his will, signed in 1799, appeared two covering jacks and three young ones, ten she asses, forty-two working mules, and fifteen younger ones.

It is interesting, however, to know that Washington appreciated thoroughly all the qualities which render mules so valuable as agricultural animals, qualities which have rendered them appreciated in almost all civilised countries, except Great Britain. The General dilates on the stout bones of Royal Gift, which are points that are looked at by all experienced mule breeders; and he descants, rightly enough, on the "great strength of mules, on their longevity, hardiness, and cheap support, which gives them a preference of horses that is scarcely to be imagined."

At the present time numbers of Spanish and other Jacks are annually imported into the United States for the purpose of mule breeding, as was made evident by an account of the escape of nearly one hundred Spanish Jacks recently brought to the port of Liverpool from Spain for the purpose of being exported to America. The *Liverpool Courier* of Jan. 18, 1894, informs us that they were finely made, powerful looking animals. They were brought to Liverpool to be trans-shipped to America in one of the White Star steamers. Pending the trans-shipment they escaped from their quarters during the night, and in the morning it was found that nearly the whole of them were missing, but they were apprehended by the police in Prescot-street, and placed in safety. This occurrence shows the extreme care that the Americans take in the breeding of their mules for farm and city work. At the present time, in the States a large amount of the agricultural labour is performed by these animals. To so great an extent is this the case, that in one of the illustrated posters showing the utilization of a reaping machine, no less than sixteen machines are delineated cutting a wheat field of some thousand acres, the whole of the machines being drawn by mules.

The draught mule of America is somewhat lighter than thóse that have already been described as being bred in France from the Poitou jacks. This depends on the facts that lighter jacks are used in breeding them, and that they are not bred from mares as heavy as those that are employed in France. Small jacks are not regarded in America as desirable, although mere size is not considered as a criterion of the intrinsic value for breeding purposes, greater reliance being placed on pedigree and breed. The mule is so important an animal in the States, and is bred there so carefully, that it is desirable to record the system which is adopted in its production by the best breeders. An account published by Mr. Killgore, of Plattsburg, is so instructive that it is most advantageous to reproduce the following details of mule breeding from it:—

"In the province of Catalonia, in old Spain, there exists a race of asses, bred with great care for many centuries, having been introduced into that country by the Moors at the time of their conquest of that kingdom. They are black in colour, with white or mealy muzzles, and whitish or greyish bellies, varying but little in form, but greatly in size, running from fourteen to sixteen and a half hands high. They are remarkable for their high carriage, fine hair, great muscular development, and superior action, in strong contrast with the common scrub donkey of the States.

"Before the late civil war these jacks were imported into Charleston, South Carolina, and were thence distributed throughout the mule-growing region of the United States. They made their mark wherever tested, showing as much improvement in mules as in any other department of live stock.

"They developed one very marked peculiarity, and that was the uniform, strong colour, good shape, fine, thrifty growing, and feeding qualities and docile temper of the mules produced from every quality and colour of dam; and, notwithstanding their variation in size from fourteen to sixteen hands high, any given mare would produce as large and fine finished and valuable a mule from the fourteen hand jack as from the sixteen hand one, thus proving the uniformity of their breeding, and showing the variation in size of their mules to be owing to the influence of the dam. A finely formed, high carried, good boned Catalonian jack, fourteen and a half hands high, is of more value for breeding mules than a sixteen and a half hand Kentucky jack. Prior to the introduction of the Catalonian blood into Kentucky, the jacks in use were mere donkeys, selected for their size, and perfectly devoid of quality, and the mule of that day had neither size, action, nor carriage, except where he chanced to be bred from a blood mare, hence blood mares were sought for as mule breeders. Now when the breeder has secured a blood jack, cold blooded mares are found to produce fine, gay, active, high priced mules; yet, even now, the more blood in the dam, the more valuable the mule. The finest mule I ever saw was by a pure Catalan jack, fourteen hands, and from a dam fifteen hands high, bred from an imported Yorkshire sire.

"The first pure-blooded Catalan jack ever brought to Kentucky was in 1832 by the Hon. Henry Clay. His sire and dam had been imported from Spain into Maryland, where Mammoth Warrior was foaled. Warrior, as he was called, was fifteen hands high. Kentucky at that time had no jennies (female donkeys), but mongrels, mostly a light shade of blue, with grey, buff, and grizzly hair, nearly as stiff as hog bristles, generally with a coloured stripe across the shoulders and down the back, ewe necked, flat in the rib, low carriage, and heavy headed, entirely destitute of any good quality except hardihood and ability to get a living where any other animal save a goat would have starved to death. With such jennies began the first effort to improve the race in Kentucky, and to Warrior they flocked in droves. He seemed to cross advantageously with them, just as the Cashmere goat crosses on the common hairy goat. His progeny seemed rapidly to lose the leading traits of their dams, and to inherit in a remarkable degree the colour and outward characteristics of their sire. Four years thereafter, Dr. Davis, of South Carolina, imported direct from Spain the second pure jack, Mammoth by name, sixteen hands high, and of great weight to his height. To Mammoth was mated the young Warrior jennies, then just maturing, thus making the second cross of pure blood, and upon these two crosses rest to-day the breeding of the race of jacks known throughout the United States as the Kentucky jack. It will thus readily be seen that Kentucky owes her position and character as a mule and jack-breeding State to this direct infusion of Catalan blood. In fact, I risk nothing when I attest that no jack in America has acquired celebrity as a mule breeder unless more or less partaking of Catalan blood, and that there is not one large, smooth, active mule on this continent not indebted to the same infusion of this potent and powerful blood.

"Sixteen hand specimens are not uncommon among the descendants of Mammoth Warrior and Mammoth, nearly all Kentucky jacks, uniting the blood of both, with many others running down to fourteen, fourteen and a half, and fifteen hands high, but they are all mongrels, being almost universally bred from jennies devoid of any breeding; this accounts for the fact that an imported Catalan jack fourteen and a half hands high is fully equal as a mule breeder to the sixteen hand native.

"The writer has seen the test fairly made time and again. He once owned a cold-blooded, open, large breeding mare, sixteen hands high. He bred her repeatedly to a Catalan jack, fourteen hands high, producing strictly first-class mules, and he afterwards, for the sake of the experiment, bred her to Mammoth, the imported jack alluded to, and bred a mule every way inferior to her general breeding from the smaller jack. The union of jack and mare sixteen hands produced a mule of even greater height than either; but leggy, light bodied, and light chested, and every way undesirable.

"Tall jacks and tall mares will never produce mules the equal of tall mares and heavy jacks from fourteen and a half to fifteen hands high. In fact, sixteen hand jacks almost invariably lack shape, action, muscle, and are generally weak constitutioned, and are not calculated to breed really serviceable mules.

"The tendency has been to breed for mere height, which is a great blunder, and should be abandoned, and more attention paid to weight, action, high quality, and purity of blood."

One of the most instructive papers on the utilisation and breeding of mules in America was published in the Report of the Commissioners of Agriculture, that was presented to the House of Representatives, in 1863. It was written by Mr. J. T. Warder, of Springfield, Ohio, himself a large breeder of mules. From this valuable Report the following extracts are taken:

"The mule is everywhere hardier than the horse, subject to fewer diseases, more patient, better adapted to travelling on rugged and trackless surfaces, less fastidious as to its food, and much less expensive in feeding, more muscular in proportion to its weight, and usually living and working to about double the age.

"In our own country the prejudice that once existed against them is rapidly yielding, and we find them used in the street cars in some of our cities, and occasionally observe them attached to elegant private carriages. In many parts of the country they are used for heavy draught; for this purpose they have long been employed in some of the iron regions, which are often hilly, and even mountainous, and traversed with very bad roads—rough, rocky, and muddy—where these animals are found to be better adapted to the circumstances than horses. In some of the mountainous portions of Pennsylvania they are used in the log-waggons, and it is truly marvellous to see them tugging at their loads, drawing the wains around huge rocks, logs, and stumps, and through rapid torrents, and among thickets of tangled underbrush that would appal a team of horses, and where these latter animals would be entirely worthless. It is true the teams employed in such situations are of superior quality, and are much larger and heavier than common mules; but their powers of endurance and their determined pluck and perseverance in overcoming difficulties make them invaluable in this kind of service. Then, again, their great intelligence adds to their value in the wild roads they have to traverse, and enables their driver to manage them without a line, but simply by the word of command.

"In the army service mules have been very extensively employed, and increasingly so within a few years. The teams consist of four and six animals, which are found to draw as much as horses, to be more easily maintained, and to endure more hardships.

"In England, where the donkeys are the property of the poor, and are considered of little value, and where the poorer mares are used for crossing, the resulting mule is an inferior animal, and is employed in very subordinate situations.

"In the mule we have the size and activity of the horse, combined with the form and hardihood of the ass, while he surpasses both his parents in sure-footedness and in longevity, and has more endurance and greater power of recuperation from fatigue and exhaustion when excessively worked. Well-bred mules are as spirited, and equally active, or even quicker than horses, if perfectly broken. They will walk fast, and in the draught they pull even more steadily. Their intelligence is so great that they may be trained very readily either to the line or to the word, and many splendid, large teams are driven, even over rough ground where there is scarcely any road, perfectly guided by the voice of the teamster.

"In the production of mules for Government use the jack should be from 14 to 15 hands high, with a good length of body, depth of chest, and with a round barrel, as indications of a good constitution. He should have heavy, flat-boned limbs, a long, thin face, with fine, thin under jaw-bones. His ears should be carried upright, and they must not be too thick. The animal should have a sprightly temper and appearance, as these qualities will almost always be transmitted to his progeny.

"The jack must be fed with a view to the maintenance of the greatest physical vigour, so as to produce an even lot of colts, and to this end he should rarely be allowed to serve more than fifty mares during the season of three months. He should be provided with such food as will give him strength without inducing feverishness. Natural exercise, with the freedom of a grass lot, should always be allowed, when practicable. Animals designed for crossing with mares should be kept from any intercourse with their own kind, as they often become entirely useless for cross-breeding when allowed contact with their own species.

"Whether it arise from a greatly-increased demand for these beasts in our country, which is now swollen by the enlarged wants of the army and its immense transportation, or whether it has come from a higher appreciation of the mule, it is certain that the number produced at the present time is vastly greater than at any former period of our history. Some shrewd agriculturist may have made the discovery that it costs less to breed and raise a mule to a suitable size than a horse; that less time is required to prepare a lot of mules than a lot of colts for the market; that young mules may be sold readily at any period, and in any amount; and more than this, that they uniformly command a higher price than a drove of horse colts of similar relative quality and value. Moreover, it may have become apparent that mules are subject to fewer diseases, that they are less liable to serious accidents, and that they are altogether more certain of producing satisfactory results from their production than horses. All of which may be set down as well-established axioms. The fact remains (whether explained or not is immaterial) that the mules of the United States have greatly increased in numbers.

"The census tables show that the number of mules produced has increased in a greater ratio than those of any other kind of farm stock, and that from 1850 to 1860 the total number of these animals had more than doubled."

The most complete as well as the most recent article on the mule, as utilised in the United States, has been published in the last volume of the Annual Reports of the Bureau of Animal Industry, printed by order of the Senate. This account is exceedingly exhaustive, and of a most practical and useful character, so much so that it has been thought desirable to reproduce it *in extenso,* with the exception of a few references and paragraphs applicable to mule rearers in the States, and the omission of those points that have previously been discussed in this volume. The article is entitled as follows:

"THE MULE.

"By J. L. Jones, Columbia, Tennessee.

"There are two kinds or classes of the mule, viz., one the produce of the male ass or jack and the mare; and the other, the offspring of the stallion and female ass. The cross between the jack and the mare is properly called the mule, while the other, the produce of the stallion and female ass, is designated a hinny. The mule is the more valuable animal of the two, having more size, style, finish, bone, and, in fact, all the requisites which make that animal so much prized as a useful burden-bearing animal. The hinny is small in size, and is wanting in the qualities requisite to a great draught animal. This hybrid is supposed not to breed, as no instance is known to us in which a stallion mule has been prolific, although he seems to be physically perfect, and shows great fondness for the female, and serves readily. There are instances on record where the female has produced a foal, but these are rare.*

"The mule partakes of the several characteristics of both its parents, having the head, ear, foot, and bone of the jack, while in height and body it follows the mare. It has the voice of neither, but is between the two, and more nearly resembles the jack. It possesses the patience, endurance, and sure-footedness of the jack, and the vigour, strength, and courage of the horse. It is easily kept, very hardy, and no path is too precipitous or mountain trail too difficult for one of them with its burden. The mule enjoys comparative immunity from disease, and lives to a comparatively great age. The writer knows of a mule in Middle Tennessee that, when young, was a beautiful dapple gray, but is now thirty years old, and is as white as snow. This

* These are probably examples of induced lactation as described in Chapter XII.

mule is so faithful and true, and has broken so many young things to work by his side, that he bears the name of 'Counsellor.' The last time he was seen by the writer he was in a team attached to a reaper, drawing at a rate sufficient to cut fifteen acres of grain per day.

"At this day mules are used extensively in nearly all parts of the country where agricultural pursuits are carried on, as well as in the mining regions, the cotton belt, and all sugar-growing countries, where they have largely supplanted the horse, and are prized highly for their gentleness and faithfulness.

"In the United States the principal States in which mules are raised are as follows, in their order as to numbers foaled in 1889 viz., Missouri, 34,500; Texas, 25,300; Tennessee, 19,500; Kentucky, 18,200; Kansas, 8200; Arkansas, 6600; Illinois, 6400; California, 5000; Indiana, 4400; Mississippi, 4200; Alabama, 3500; North Carolina, 3300; Iowa, 2300; Nebraska, 2300; Georgia, 2000; Virginia, 2000; Louisiana. 1300; Oregon, 1300; Ohio, 900; South Carolina, 700; and Pennsylvania, 600. Many other States raised mules, making the number foaled, in 1889, 157,000.

"Kentucky mules are showy, upheaded, fine-haired animals, their extra qualities being attributable to the strong, thoroughbred blood in the greater part of their dams. The same may be said of Tennessee, where it is thought the climatic influences produce a little better, smoother, and finer hair, coupled with early maturity, which qualities are much prized by an expert buyer.

"The mules in Missouri, Illinois, Indiana, and some other of the so-called North-western States, have large bone, foot, body, and substance, and possess great strength, but they are wanting in that high style, finish, and fine hair that characterise the produce of some of the States further south, and are longer in maturing. Mule breeding in these States is one of the most important branches of industry, and is supposed to date back prior to 1787.

"There is no kind of labour to which a horse can be put for which a mule may not be made to answer, while there are many for which mules are more peculiarly adapted than horses; and among the rest, that of mining, where the mule is used, and many of them need no drivers. They can endure more hardships than the horse, can live on less, and do more work on the same feed than any other beast of burden we use in America.

"A cotton-planter in the South would feel unwilling to raise his crop with horses for motive power. The horse and the labour of the cotton belt could not harmonise, while the negro is at home with the mule.

"A mule may be worked until completely fagged, when a good feed and a night's rest will enable it to go; but it is not so with a horse.

"The mule being better adapted for carrying burdens, for the plough, the waggon, building of railroads, and in fact all classes of heavy labour, let us see how it compares with that noble animal, the horse, in cost of maintenance.

"From repeated experiments that have come under my observation in the past twenty-five years, I have found that three mules, 15 hands high, that were constantly worked, consumed about as much forage as two ordinary-sized horses worked in the same way, and while the mules were fat the horses were only in good working order. Although a mule will live and work on very low fare, he also responds as quickly as any animal to good feed and kind treatment. True, it is charged that the mule is vicious, stubborn, and slow, but an experience in handling many mules on the farm has failed to sustain the charge, save in few instances, and in these the propensities were brought about by bad handling. They are truer pullers than the horse, and move more quickly under the load. Their hearing and vision are better than the horse. The writer has used them in all the different branches of farming, from the plough to the carriage and buggy, and thinks they are less liable to become frightened and start suddenly; and if they do start, they usually stop before damage is done, while the horse seldom stops until completely freed. The mule is more steady while at work than the horse, and is not so liable to become exhausted, and often becomes so well instructed as to need neither driver nor lines.

"In the town in which the writer lives, a cotton merchant, who is also in the grocery trade, owned a large sorrel mule, 16 hands high, that he worked to a dray to haul goods and cotton to the depot, half a mile from his business house. This mule often went the route alone, and was never known to strike anything, and what was more remarkable, would back up at the proper place with the load, there being one place to unload groceries and another for cotton.

"They are also good for light harness, many of them being very useful buggy animals, travelling a day's journey equal to some horses. The writer obtained one from a firm of jack breeders in his vicinity, that was bred by them, as an experiment, being out of a thoroughbred mare by a royally bred jack. She is 16 hands high, as courageous as most any horse. In travelling a distance of thirty-two miles, this mule, with two men and the baggage, made it, as the saying goes, 'under a pull,' in four hours, and when arrived at the journey's end seemed willing to go on.

"We do not wish to be understood as underrating the horse, for it is a noble animal, well suited for man's wants, but for burden-bearing and drudgery is more than equalled by the patient, faithful, hardy mule.

"THE KIND OF SIRE TO BREED FROM.

"There are two kinds of jacks—the mule breeding and the ass breeding jack*, the latter being used chiefly in breeding jacks for stock purposes. It is only with the mule breeding jack that we will deal.

"A good mule jack ought to be not less than 15 hands high, and have all of the weight, head, ear, foot, bone, and length that can be obtained, coupled with a broad chest, wide hips, and with

* The term jennet is used in the United States to signify a female ass, and hence the jacks employed in breeding "jack stock" are termed "jennet jacks." The best jacks are usually selected for this purpose, and command a service fee of $50 (10*l*.).

all the style attainable with these qualities. Smaller jacks are often fine breeders, and produce some of our best mules, and when bred to the heavier, larger class of mares show good results, but as 'like produces like,' the larger jacks are preferable.

"Black, with light points, is the favourite colour for a jack, but many of our grey, blue, and even white jacks have produced good mules. In fact, some of the nicest, smoothest, red-sorrel mules have been the product of these off-coloured jacks; but the black jacks get the largest proportion of good-coloured colts from all coloured mares.

"The breed of the jack is also to be looked into. There are now so many varieties of jacks in the United States, all of which have merits, that it will be well to examine and see what jack has shown the best results. We have the Catalonian, the Andalusian, the Maltese, the Majorca, the Italian, and the Poitou—all of which are imported—and the native jack. Of all the imported, the Catalonian is the finest type of animal, being a good black, with white points, of fine style and action, and from 14½ to 15 hands high, rarely 16 hands, with a clean bone. The Andalusian is about the same type of jack as the Catalonian, having perhaps a little more weight and bone, but are all off colours. The Maltese is smaller than the Catalonian, rarely being over 14½ hands high, but is nice and smooth. The Majorca is the largest of the imported jacks, the heaviest in weight, bone, head, and ear, and frequently grows to 16 hands. These are raised in the rich island of Majorca, in the Mediterranean Sea. While they excel in weight and size, they lack in style, finish, and action. The Italian is the smallest of all the imported jacks, being usually from 13 to 14 hands high, but having good foot, bone, and weight, and some of them make good breeders. The Poitou is the latest importation of the jack, and is little known in the United States. He is imported from France, and is reported to be the sire of some of the finest mules in his native land. These jacks have long hair about the neck, ears, and legs, and are in some respects to the jack race what the Clydesdale is to other horses. He is heavy set, has good foot and bone, fine head and ear, and of good size, being about 15 hands high.

"The native jack, as a class, is heavier in body, having a larger bone and foot than the imported, and shows in his entire make-up the result of the limestone soil and grasses common in this country. He is of all colours, having descended from all the breeds of imported jacks. But the breeders of this country, seeing the fancy of their customers for the black jack with light points, have discarded all other colours in selecting their jacks, and the consequence is that a large proportion of the jacks in the stud now, for mares, are of this colour.

"The native jack, being acclimated, seems to give better satisfaction to breeders of mules than any other kind. From observation and experience it is believed that our native jacks, with good imported crosses behind them, will sire the mules best suited to the wants of those who use them in this country, and will supply the market with what is desired by the dealers. The colts by this class of jacks are stronger in make-up, having better body, with more length, larger head and ear, more foot and bone, combined with style equal to the colts of the imported jacks.

"While many fine mules are sired by imported jacks, this is not to be understood as meaning that imported jacks do not get good foals, yet, taken as a class, we think that the mule by the native jack is superior to any other class. This conclusion is borne out by an experience and observation of some years, and by many of the best breeders and dealers in the United States.

"THE KIND OF MARE TO BREED FROM.

"As the mule partakes very largely in its body and shape of its mother, it is necessary that care should be taken in selecting the dam. Many suppose that when a mare becomes diseased and unfit for breeding to the horse, then she is fit to breed for mules. This is a sad mistake, for a good, growing, sound colt must have good, sound sire and dam.

"The jack may be ever so good, yet the result will be a disappointment unless the mare is good, sound, and properly built for breeding. First, she should be sound and of good colour; black, bay, brown, or chestnut is preferred. Her good colour is needed to help to give the foals proper colour, and this is a matter of no small importance.

"This should not be understood as ignoring the other colours, for some of the best mules ever seen were the produce of grey or light-coloured mares, as many dealers and breeders will attest. The mare should be well bred; that is, she would give better results by having some good crosses. By all means let her have a cross of thoroughbred, say one-quarter, supplemented with strong crosses of some of the larger breeds, and the balance of the breeding may be made up of the better class of the native stock. The mare should have good length, large, well-rounded barrel, good head, long neck, good, broad, flat bone, broad chest, wide between the hips, and good style.

"HOW TO BREED THE MULE.

"Having selected the sire and the dam, the next thing is to produce the colt. The sire, if well kept and in good condition, is ready for business, but not so with the mare. The dam is to be in season; that is, in heat. Before being bred, to prevent accidents, the mare should be hobbled or pitted. Having taken this precaution, the jack may be brought out, and both will be ready for service. Care should be taken not to overserve the jack, as he should not be allowed to serve over two mares a day.

"The mare, after being served, may be put to light work, or put upon some quiet pasture by herself for several days until she passes out of season, when she may be turned out with other stock to run until the eighteenth day, when she should be taken up to be teased by a horse, to ascertain if she be in season, and if so, she should be bred again. Some breeders think the ninth, some the twelfth, and some the fifteenth day after service is the proper day to tease, but observation has taught us that the best results come from the eighteenth-day plan. After she becomes impregnated she should have good treatment; light work will not hurt her, but care should be taken not to overexert. She should have good, nutritious grass if she runs out and is not worked, but if worked she should be well fed on good feed. The foal will be due in about 333 days. As the time approaches for foaling the mare should be put in a quiet place, away from other stock, until the foal is dropped. She will not need any extra attention, as a rule, but should be looked after to see that everything goes right.

"After the foal comes it will not hurt the mare or colt for the dam to do light work, provided she is well fed on good, nutritious food. Should she not be worked and is on good grass, and fed lightly on grain, the colt will grow finely, if the mare gives plenty of milk; if she does not the foal should be taught to eat such feed as is most suitable.

"The colt should be well cared for at all times, and particularly while following its mother, for the owner may want to sell at weaning time, which is four months old, and its inches then will fix the price. Good mules at weaning time usually bring from $75 to $90, and sometimes as high as $100 (18*l.* to 25*l.*).

"Feeders, dealers, and buyers prefer the mare mule to the horse, and they sell more readily. The females mature earlier, are plumper and rounder of body, and fatten more readily than the male.

"In weaning the colt, much is accomplished by proper treatment preparatory to this trying event in the mule's life. It should be taught to eat while following its mother, so that when weaned it will at once know how to subsist on that which is fed to it. The best way to wean is to take several colts and place them in a close barn, with plenty of good, soft feed, such as bran and oats mixed, plenty of sound, sweet hay, and, in season, cut-grass, remembering at all times that nothing can make up for want of pure water in the stable. Many may be weaned together properly. After they have remained in the stable for several days they may be turned on good, rich pasture. Do not forget to feed, as this is a trying time. The change from a milk to a dry diet is severe on the colt. They may all be huddled in a barn together, as they seldom hurt each other. Good, rich clover pastures are fine for mules at this age, but if they are to be extra fine, feed them a little grain all the while.

"There is little variety in the feed until the mules are two years old, at which time they are very easily broken. If halter-broken as they grow up, all there is to do in breaking one is to put on a harness and place the young animal beside a broken mule, and go to work. When it is thoroughly used to the harness the mule is already broken. Light work in the spring, when the mule is two years old, will do no hurt, but, in the opinion of many breeders and dealers, make it better, provided it is carefully handled and fed.

"HOW TO FATTEN THE MULE.

"This is one of the most important parts of mule-raising, for when the mule is offered to a buyer, he will at once ask, "Is he fat?' and fat goes far in effecting a sale. A rough, poor mule could hardly be sold, while if it is fat the buyer will take it because it is fat.

"The mule should be placed in the barn with plenty of room, and not much light, about the 1st of November, before it is two years old, and fed about twelve ears of (Indian) corn per day, and all the nice, well-cured clover hay it will eat, and there kept until about the 1st of April. Then in the climate of middle Tennessee the clover is good, and the mule may be turned out on it, and the corn increased to about twenty ears or more per day. They will then eat more grain, without fear of 'firing;' that is, heating so as to cause scratches, as the green clover removes all danger from this source. During the time they run on the clover they eat less hay, but this should always be kept by them. About the 1st of May the clover blooms, and is large enough to cut, in the latitude of Tennessee. The mules should be placed, then, in the barn, with a nice smooth lot attached, and plenty of pure water. A manger should be built in the lot, 4ft. wide by 4ft. high, and long enough to accommodate the number of mules it is desired to feed. This should be covered over by a shed high enough for the mule to stand under, to prevent the clover from wilting. The clover should be cut while the dew is on, as this preserves the aroma, and they like it better. While this is going on in the lot, the troughs and racks in the barns should be supplied with all the shelled corn (maize) the mules will eat. 'Why shell it?' some one will ask. Because they eat more of it, and relish it. A valuable addition at all times consists of either short-cut sheaf oats, or shelled oats, and bran, if not too expensive.

"From this time the mule should be pressed with all the richest of feed, if it is desired to make it what is termed in mule parlance, 'hog fat.' Ground barley, shelled oats, bran, and shelled corn, should be given, not forgetting to salt regularly all the while, nor omitting the hay and green corn blades. While all those are essential, oats and bran, although at some places expensive, are regarded as the *ne plus ultra* for fattening a mule, and giving a fine suit of hair. Be sure to keep the barn well bedded, for if the hair becomes soiled from rolling it lowers the value, as the mule is much estimated for its fine coat.

"The grain makes the flesh, and the green stuff keeps the system of the mule cool, and balances the excess of carbonaceous elements in the grain fed.

"The manner of feeding, if properly carried out, with the proper foundation to start with, will make mules, two years old past, weigh from 1150lb. to 1350lb. by the 1st of September, at which time the market opens.

"A feeder of eighteen years' experience claims that oats and bran will put on more fine flesh in a given time, coupled with a smoother, glossier coat of hair, than any other known feed. The experienced feeder follows this method from weaning till two years old."

The endurance and utility of the American mule was thoroughly demonstrated during the Civil War, when a large number of these animals performed extraordinary service in connection with the Federal armies. One six-mule team fitted out in Maryland in the spring of 1861, driven by a coloured driver, was worked in Washington until May, 1862, then transferred to the army of the Potomac, re-shipped for Washington, employed in hauling ammunition at the battle of Bull Run, and afterwards followed the army of the Potomac under Grant till 1864. The mules were worked every day until Richmond was taken, and in 1865 transferred back to Washington, and at the end of 1866 were still working in the train, and regarded as one of the best military teams going. They were all under 14½ hands. They had frequently been without bite of hay or grain for four or five days, and for twenty-four hours without water.

Riley, in his work on the mule, published in 1867, gives numerous examples of other teams which did equally good work.

CHAPTER XVI.

MULES FOR MILITARY SERVICE.

The advantages of mules as pack animals for military and draught purposes are acknowledged on all hands. Whenever any branch of the army is employed on foreign service mules have to be purchased for transport, inasmuch as horses cannot stand the rough labour that is required of them.

No stronger or more conclusive testimony as to the invaluable service rendered by mules when employed for army transport can be adduced than that furnished by Major A. G. Leonard, in his recently published admirable book on "The Camel," considered solely as an animal for military use. Although the author is writing a work on one animal and detailing the advantages that it offers to the military service when employed in suitable situations, his experience of mules, of which he has had four hundred at one time under his command, leads him to express himself in the strongest possible manner in their favour. Major A. G. Leonard writes:

"The mule is about the handiest and hardiest of all pack animals. He can work in any country, and under every condition of climate, but is specially suited for mountainous regions. He will go over any ground, no matter how steep and rocky, he is so very sure-footed and nimble. His toughness and endurance are perfectly marvellous, and it is wonderful how long and on how little he seems to live and even thrive. He is less liable to sore backs and galls than any other animal, the donkey excepted. He is a fast walker, and will keep up three miles an hour on average ground, and on good I have known him to do three and a half. Even on a bad road, over rocks and hills, he will do two and a half miles; but of course heavy sand is very trying to him, as it is for all animals except the camel. He is accused of being obstinate and ill-tempered, but this—if it is the case—arises almost wholly from ill-treatment during juvenility, as well as from the woeful ignorance of the animal's ways that generally prevails among Britishers. The mule is naturally docile and patient in the hands of those who understand him and who treat him kindly, and he will show them as much affection nearly as a horse. He strongly objects to be hit over the head and kicked violently in the ribs or stomach, as I have frequently caught "Tommy Atkins" doing, and naturally enough this brutal treatment by no means improves his temper or his manners, so he returns it by biting, kicking, and becoming generally refractory. It is generally supposed that they live from fifteen to twenty years, though some live to thirty, and a few beyond that age. When I was in India fifteen years ago there were mules belonging to the Commissariat who were said to have been twenty-two years in the service, and were still working.

"The Indian pack mule, or I should say the pack mule used in India, ranging between 12 and 13 hands, is by far the best I have seen. I dislike taller mules for pack work. The shorter ones are handier and much easier to load, much more so when they are fresh and obstreperous, as at the beginning of a march or after a rest. In the Egyptian Expedition of 1882 I worked with four hundred Sicilian mules, and splendid animals they were too, but, on the whole, they were in my opinion a trifle too tall for pack work."

Another very practical authority on military transport, Captain F. D. Lugard, in his work on our East African Empire, writes as follows:

Gun Mule. (Height 13.2. Indian Mountain Battery.)

"Of all transport animals the hardiest, and therefore, on the whole, perhaps the most useful, is the mule. To be worth his keep and supervision, mules should not be less than 13 hands high, and capable of carrying 180lb. to 200lb. over rough country. This they will do if provided with a suitable saddle, so that the load may ride easily, and sore backs and continual breakdowns may be avoided."

The engraving shows a gun mule in marching order as used in the Indian Mule Batteries, but, as Captain Lugard says, the character of the saddle is most important, and therefore it is desirable to reproduce the following detailed description of the best pack saddles used in India, the native one which is used in the Punjab and the Government gear which has been founded on it. This very important detailed account was furnished by a military officer in an exhaustive article on Indian transport animals, communicated to the *Times*, September 21st, 1880. In the course of his introductory remarks, the writer says:

"The mule is, probably, the best of all transport pack animals for a mountainous country. Unfortunately, she—for the female is generally employed—is comparatively scarce in India. The mules purchased for transport purposes in Afghanistan were, generally speaking, small-sized animals, seldom exceeding 13 hands, but they have wonderful powers of endurance, and are seldom 'sick or sorry.' For some time past the Government of India have endeavoured to improve the breed of mules in the North-Western Provinces, and it is satisfactory to learn that the question of the best means of extending mule breeding throughout India is now engaging serious attention.

"The nature of pack saddle in general use for mules and ponies is that known as the 'Punjab pattern.' It is a modification of the native pack-gear which is met with all along the North-Western frontier of India. The native mule-owner, when 'saddling up' or preparing the animal to receive the load, commences by laying several strips of a soft material, such as old woollen blanketing, along each side of the animal's backbone, extending from the withers to the quarters. The country name for this padding is 'malli.' The horse blanket, or 'jhool,' as it is sometimes called—'jhool' being the Indian name for the clothing of all transport animals—is then folded double or quadruple, according to its size, and placed on the animal's back over the 'malli.' The 'soonda' is now placed on the blanket, and the whole gear bound tightly on to the animal by the 'dotunga.' The object of the 'soonda' is to keep the load from pressing on the animal's spine, and to distribute the pressure evenly along the back. It is made by stuffing with reeds or stout straw a long bag made of canvas or blanketing. This bag, which is like a six-foot sausage, and as thick as an ordinary wine bottle, is then bent into the shape of a cylindrical sugar-tongs, the legs of which are kept from splaying out by 'keepers' of canvas or blanketing. When placed in position on the animal's back, the bent end is a little in rear of the withers, and the spine is between the two legs. The 'dotunga' is simply a canvas cover or body roller, fitted with girths, and sometimes with breast and breech pieces; the latter, however. are usually made of tape or string. The 'dotunga' is placed over the 'soonda' in the centre of the back, and when girthed up, binds all the gear together. The load is now slung over the animal and lashed on. The Government gear mainly consists in replacing the canvas 'dotunga' by a species of saddle made of two well-stuffed leather flaps, fastened by leather bands, and fitted with strong girths, crupper, and leather breast and breech pieces. The saddle is placed so that the leather bands rest on the 'soonda,' and the stuffed flaps protect the sides of the animal. Two iron rings are fitted to each flap, so that the load can be firmly attached to the saddle. The load is carried either in a 'sulletah' or a 'sling,' or simply lashed on. The transport 'sulletah' is a double bag, made of coarse canvas, sacking, or cloth. Its size depends on the animal it is required for—elephant, camel, or mule. It may easily be made by folding a broad strip of material until the ends meet in the centre, or, rather, until one end slightly overlaps the other. The sides are then sewn together and the ends furnished with strings or tapes. The result is a double bag or purse, which can be filled on each side, and slung across the animal's back. Some mule 'sulletahs' are specially fitted with leather thongs, for attachment to the iron rings of the Punjab saddle. The 'sulletah' is useful for carrying small packages, loose grain, articles packed in thin coverings, &c. The 'sling' is a broad strip of coarse canvas or sacking furnished along its edges with eyelet holes. The load, which may consist of boxes, portmanteaux, sacks of flour, &c., is arranged on the sling so that the ends can be folded up and the sides lashed together by a cord running through the eyelet holes. Thus a purse with open sides can be formed and slung across the animal. The size of the 'sling,' of course, depends upon the nature of the animal."

A most useful and exceedingly interesting account of the manner in which baggage can be securely fastened on to pack mules without the aid of any special saddle was described and illustrated in the *Field* of February 2nd, 1895, by Mr. Albert H. Leith, of Chihuahua, Mexico. The account, with some slight alteration in the text and engravings, is as follows. Mr. Leith says:

"I was initiated into the mysteries of the hitch, by means of which baggage is securely fastened on the most refractory of ponies, during the campaign in Afghanistan, and I was much impressed on seeing how neatly and securely the load was tied by means of this knot, which is in use over all the Pacific slope; and calling to mind the scenes I had sometimes witnessed, I thought that the accomplishment would be an exceedingly useful one to the British soldier.

"Twenty years of frontier life and use of the hitch enable me to thoroughly realise its advantages, and having frequently seen allusions to it, I think that perhaps the endeavour to explain it may be acceptable. Its greatest advantage is that in the case of camp outfits, when blankets are a part of the pack, no saddle is required; indeed, the pack is infinitely firmer, and the animal less liable to be given a sore back, without the forward shifting abomination which the pack saddle is.

"In both of the diagrams the off side of the mule only is shown. To proceed:

"Take a thirty-foot picket rope, throw half on each side of the horse or mule, the middle of the rope lying across the top of the pack; then let each man make a loop round the pack on his side, putting his foot into it as a stirrup, as shown in the figure. Then the man on the off side takes his end of the rope as shown in the drawing, and, passing it down through his stirrup-loop, puts it under the belly and through his companion's stirrup-loop on the other side (both meanwhile holding taut with one hand above). When he has pulled the slack of his part of the rope through (but not till then) he tells his companion to slip his foot out, and at the same time smartly hauls the caught-up stirrup-loop from the near side into its place

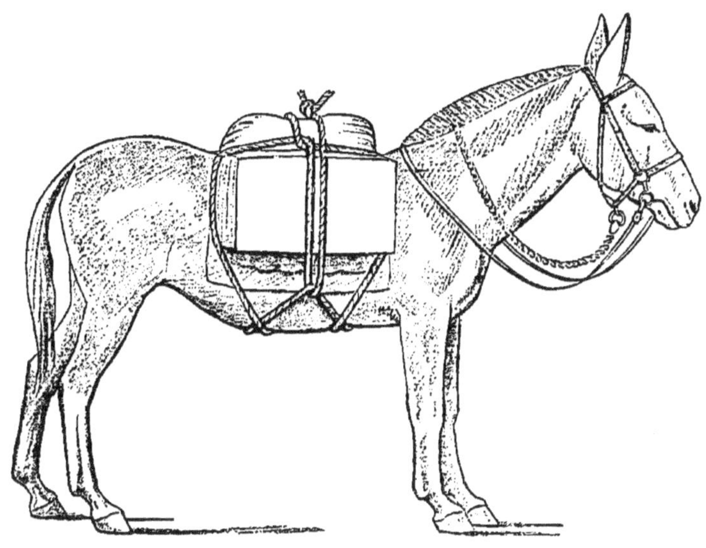

under the mule's belly, as shown in the second drawing. Then his companion in turn takes his end of the rope and, reaching under the belly, puts it through the stirrup-loop on the off side (which the first man has kept his foot in) and hauls it similarly into place on the near side of the belly, then both on their respective sides, giving a good pull together, make everything taut (as shown in the drawing), and all that remains to be done is to tie the spare rope ends with a double-reef knot (pulling tight again when making it) on the top of the pack.

"Now, with reference to what I said above about pack saddles. All the saddle that this tie requires is a large pad; therefore if blankets are a part of the pack, they make the pad. First lay an old half blanket as a sweat-cloth on the animal's back (folded so as to cover about two and a half feet length of the back, and hanging down a little more than half-way down the ribs); then folding all the blankets and bedding to the same size, as much as possible, lay them on top of the sweat-cloth, evenly one by one; on the top of that lay the canvas or waterproof sheeting, similarly folded; then sling flour and other provisions, in sacks, equally balanced on each side of back (by means of small rope ties connecting them and holding them in place); then put whatever other sack of dunnage there is still to go on, on top in the middle between the two last; and then, over all, holding everything together, goes the hitch. And if this is carefully put on as regards balancing of weights, and made well taut in all its parts, it will 'stick' over the roughest mountain trails, and when the pack is taken off at night there will be no sore back, as is so frequently the case with a pack saddle.

"I have used this hitch under all circumstances, having packed only 20lb. of blankets with it on a spare horse when going on a cattle round-up, or 200lb. of general camp outfit on a mule when crossing mountain trails where a wagon could not go. It is too well known in the Far West to require any testimonials, but one, I think, I may give it. Twelve years ago, when I settled the ranche on the Mexican frontier from which I write, smuggling was the occupation of the Mexicans in the frontier villages, and one day one of the smugglers, who had done me some favours (*honi soit qui mal y pense*), camped with his mule train in the mountains at a place where I was 'nooning.' Well, Mexicans are conceded to be good packers, and especially the mountain smugglers, but they use a more complicated tie than this hitch, so I taught it him. From that day till smuggling was put an end to by an efficient force of frontier gendarmes, he used no other, and showed it to many of his *confrères*, the consequence being that to-day it is known in the neighbouring Mexican villages as the *nudo contrabandisto*."

In India mule breeding for the army service has attracted very great attention. It was followed in the Punjab before the country came under British rule, as the mule was found an indispensable animal for traffic over the mountain passes in the north-west of the country. Since 1876 the Government, according to the report of Vet.-Lieut.-Colonel J. H. B. Hallen, C.I.E., General Superintendent Horse-breeding Operations, has fostered the mule-breeding industry by giving prizes for the best mules and mule-breeding stock, and by utilising the best donkey sires obtainable. Colonel Hallen states that at the annual fair held at Rawalpindi, from thirteen to fourteen hundred mules are as a rule exhibited for sale, and that owing to the employment of good sires they are improving year by year. At first the natives refused to breed mules, from some superstitious feeling, but finding that they fetched a much larger price at the fairs than the horse stock bred from their mountain ponies, they went largely into their production. Colonel Hallen, in his official Memorandum on mule breeding, April, 1891, reports that:

"It may be noted that a very inferior mare, quite unfit for horse breeding, *i.e.*, only able to produce worthless and unsaleable horse stock, will, when mated with a good donkey, produce a mule of size and worth. Thus mule breeding has been a most useful adjunct to horse breeding, as worthless mares, totally unfit to produce horse stock, have been employed as mule breeders, and not only have given fair mules, but have started the industry in North-Western Provinces and Rajputana, and these unfit mares for horse breeding have been prevented producing bad horse stock, and so the breeding districts have been relieved of the evil influence these mares had in producing worthless horse stock, and of course their malformation and hereditary defects and diseases have been for ever got rid of, as their mule produce is infertile.

"Further it has to be explained that a good horse mare when mated with a good donkey yields, as a rule, a superior mule of great value, indeed, often of greater value in the Indian market than an ordinary horse; so in the Punjab, where mule breeding is better understood and appreciated than in other parts of India, a better class of pony and horse mare is made a mule breeder.

"At the Rawalpindi Fair of 1888 a mule realised the highest price of any stock sold at the fair. Again, as regards mule breeding being a safer investment than ordinary horse breeding, it is found that the mule is a hardier animal and able to browse young bushes, &c., so that in seasons of drought he maintains his condition and develops when horse stock become impoverished and prove starvelings."

In some of the other parts of India mule breeding had made less progress, inasmuch as it was not properly fostered by the Government and good donkey sires introduced. Colonel Hallen regards the Italian breed known as the "Razza" as the best jack for mule breeding for the Indian service, as the produce of the Poitou jack carries too much hair, and suffers in the summer months. The Catalan jacks are not as suitable as the Italian, and the Cyprians are too small and wanting in bone. The Arab jacks are found too delicate to stand the winter of the Punjab. The Persian are superior to the Arab, and as mule sires are better. Vet.-Major G. J. R. Rayment, the assistant superintendent of the Horse Breeding Department, North-Western Provinces, agrees fully with Colonel Hallen that the best class of donkey sires for his district is undoubtedly the Italian "Razza," of which a typical example is represented in the frontispiece.

During the present year a report has been received from Lieut. F. A. Thatcher on mule breeding across the Chinese frontier, bordering on the Bhamo district. He states that the extent to which the business of mule breeding is pursued in that country is almost incredible; that the produce of the mines is carried by means of mules and ponies, which are numbered by tens of thousands. Nearly one-half of the pony mares are employed to breed mules, but no hinnies are allowed to be produced. The young mules are taken in hand for training when about two years old. They do an enormous amount of work on a very small quantity of food.

The rearing of mule stock for army service in India and elsewhere is so important a matter that it has been thought desirable to add the practical directions on mule breeding, written for the use of the horse and mule breeding department in India by Mr. C. L. Sutherland, which will be found in the Appendix.

CHAPTER XVII.

REMARKS ON THE USE OF MULES.

Considering that mule breeding is increasingly carried on in most of our colonies, as well as in India, it seems rather an anomaly that the mother country should not be able to supply the "jack," the chief factor in the business. We are accustomed to supply our colonists with horses, bulls, sheep, and pigs of the very best kinds, and such as meet all requirements; but when jacks are wanted, an order is generally placed in the hands of a City firm, who at once find themselves at their wits' end in regard to carrying it out. Attempts are made to find out some mercantile house which has connection with Spain, and the order is mostly placed there in the hands of people who are not in the slightest degree acquainted with the business or its details. Animals are bought, shipped to England, and re-shipped to their destination, and on arrival, after great cost has been incurred, are found to be utterly useless for the purpose required. If the order is sent to the South of Spain, what are called by the Americans "off coloured" (grey) jacks are bought. This is the first mistake. Custom requires that mule-getting jacks shall be "black, with mealy points." The second mistake is—and this remark applies specially to Andalusian jacks—that fine-looking, big-boned animals, that the Spaniards have discovered are no good for mule breeding, are carefully palmed off on the English buyers. It is a common thing for the captains of English steamers engaged in trading with Spanish ports to bring home a jack or two on speculation on their own account; it may be taken almost as a certainty that such animals are useless, and that they have been carefully kept for this particular market. At the same time it must be owned that, apart from colour, very excellent big-boned jacks can be obtained in Southern Spain, but considerable care and technical knowledge must be exercised in making the purchases.

Seeing that our breeds of stock are so much sought after, it seems strange that, except in a few isolated instances, no attempt should have been made generally to improve the British donkey, and so give him the rank and position pertaining to a jack. It may be said that a donkey is a donkey all over the world; but the difference between a donkey and a jack is as great as that between a tramp and a King. Directly a male member of the asinine race has size, bone, and substance enough to be used as a jack, his value is increased enormously. A donkey in London is worth from 2*l.* to 5*l.* A jack (or baudet) in Poitou is worth from 100*l.* to 400*l.*; some years ago, when the writer was in Kentucky, he was assured by Mr. B. B. Groom, who will be remembered by old Shorthorn breeders, that more than one jack had been sold in Kentucky for $5000 (1000*l.*). Looking at the never-failing demand from our colonies for jacks at good prices, it might perhaps be worth the while of enterprising British agriculturists to turn their attention to the production of this class of stock.

Mules have been bred in the south of Ireland for many

years, and many foreign jacks have been imported for the purpose; but, with the exception of the late Mr. Kavanagh and the late Lord Clancarty, no attempt to breed jacks for exportation to the colonies has been made in Ireland. Some three years ago the Congested Districts Board imported several grey Andalusian jacks for breeding purposes. In the opinion of the writer, it would have been more expedient to import the "black jack, with mealy points," from, say, Catalonia. The improved Irish jack stock would in course of time have been available for exportation at good prices to the colonies for mule breeding purposes.

It will be argued, and with some degree of truth, that grey jacks of the right shape and make will get just as good mules as black. So they will; but these grey jacks will not bring the same price when offered for sale in the market as mule getters; and surely it is best to breed the kind of stock that will bring about this desirable result. A few years ago Dr. P., of Nashville, Tennessee, after trying in vain to buy some jacks in Poitou (the prices were too high) went on to the south of Spain. He was not an expert, and in his eyes a jack was a jack. He bought a number of grey "off coloured" animals, which he took over to the States for sale. Arrived in Tennessee with his cargo, he found that it was impossible to dispose of them at a remunerative price, and a heavy loss resulted.

In buying a jack for mule breeding it is requisite to decide whether the mules are required for pack or draught work, and, if the latter, whether for heavy draught or trotting work. If pack mules are wanted, a smaller and less expensive jack is required. Other things being equal, the value of a jack increases with his height. There can be no greater mistake than to employ tall mules for pack work. Indeed, it is next to impossible to load, say, a 15 hand mule, and this stupid attempt to utilise tall mules for pack work always results in failure, and then the mule is blamed. To speak generally, a pack mule should never exceed 14 hands in height. From 12 hands to 13.2 is better. The load of such an animal should not exceed 200lb., exclusive of the pack saddle, which is often unnecessarily heavy. To produce pack mules the jack may vary in height from 12 hands to 13.2, and the mare from which the mule is to be bred should be about the same height. In breeding in India it is found extremely difficult to keep down the height of pack mules. When nature allows two distinct species, like the horse and ass, to breed together, the resulting mule will generally grow to a greater height than either of its parents if properly fed during the growing stage. The Italian "Razza" jack, standing from 12 hands to 13.2, produces excellent pack mules, with great courage and endurance.

To get mules for heavy draught, whether on the farm, for the town dray, for hauling in the docks, or towing on canals, there is no jack that will surpass that of Poitou if properly chosen and mated with heavy mares. If mules are required for trotting and galloping work, as in the coaches of Southern Africa, the Catalonian jack, when he is not narrow chested and high on the leg, is the best, as he has more courage than the Poitou. Lighter and better bred mares must also be selected for this purpose, and it is essential that the mares shall be good-tempered. The Kentucky mule is well known for his courage and generally good qualities. This is chiefly on account of the mares having a cross of thoroughbred blood in their veins, and this, with the assistance of a good jack of Catalonian origin, has made the Kentucky mule what he is. The sight of a number of "smooth" Kentucky mules with shining coats and in show condition is a thing that must be seen to be appreciated. The original Maltese jack, from Gozo, formerly had a great name in the States as a mule getter, but it was stated at one time that the island had been entirely depleted of the old breed by the Americans. The Maltese jack has been much used in the West Indies, especially in the breeding "pens" in Jamaica.

At a time when many eyes are turned towards South Africa and its requirements, and when the difficulty attaching to transport in this part of the world on account of horse sickness and the tsetse fly threatens to baffle all efforts in this direction, it will be well to try and discover a solution of the difficulty, pending the construction of railroads. The standard mode of transport there, as all the world knows, is by bullock waggons; but bullocks are slow, and the rumen of the bullock takes a great deal of filling. The bullock is slow and sure, but, on account of its slowness, cannot be accepted as entirely satisfactory in these days, although it must be borne in mind that, prior to the adoption of the mule in America, the development of the Western States, so far as the transport was concerned, was entirely brought about by bullocks attached to the prairie-schooners.

In the late Lord Randolph Churchill's book on "Men, Mines, and Animals in South Africa," the donkey is declared to be exempt from horse sickness. But the donkey proper is too small for anything but pack work. Big donkeys would be too valuable, and probably too delicate, for this kind of work. At the same time we cannot afford to ignore the value of the donkey for this purpose.

Reference has already been made in these pages to the

employment of Burchell's zebra in the Transvaal coaches, and from the latest accounts he would seem to answer fairly well, as he is said to suffer from neither of these ailments, and, in addition, it has been found that the idea as to the impossibility of taming and breaking the zebra is a perfect myth. Zebras (Burchell's) have been used in harness at the Jardin d'Acclimatation and in the streets of Paris for more than twenty years. In these circumstances it would seem to be quite worth while to attempt the breeding of zebra (Burchell's) mules from mares. It might, perhaps, be said, "Why not breed them from donkey mares, inasmuch as the donkey is exempt equally with the zebra?" But mules so bred would not be fast enough for coaching work, and would take too much "getting along." Zebra hinnies would be better—*i.e.*, bred from female Burchells by a good Yorkshire hackney stallion. The hinny or jennet is always a better beast for fast work than the mule. Jennets may be seen trotting along in almost any town or village in the South of Ireland. If it is a fact that the zebra and donkey are both "exempt," it might be worth while to cross them in both ways, so as to produce both mules and hinnies, which should all be very valuable for pack work at least. These suggestions are offered for the consideration of those whom it may concern in South Africa.

Three Burchell-zebra hinnies, bred from a female Burchell by small horses, may be seen in Sir Henry Meux's park at Theobalds, near Enfield, Middlesex. The hybrids vary in height from 13 to 14 hands. They "favour" the zebra in markings and conformation, and are well worthy of inspection.

In the chapter on "The American Mule" (page 116), an exceedingly good article by Mr. J. L. Jones, of Columbia,

Tennessee, has been previously quoted almost in its entirety. Mr. Jones describes admirably the various breeds of imported jacks that are used in the States, and finishes up by declaring that the "native jacks, with good imported crosses behind them, will sire the best mules." This is entirely in accord with the writer's views. There is no one European breed that combines in itself all the desirable qualities of size, bone, short legs, and courage, but by judicious crossing of the various breeds a very superior animal can be obtained. The Poitou and Majorca have size, bone, and short legs, but are deficient in courage. The Catalonian has size, fair bone, and good courage, but is apt to be narrow chested, light barrelled, and high on the leg. The Maltese has fair height, capital courage, but is light of bone. The Italian has extraordinary courage, but is rather deficient in height, weight, and bone. So far as getting draught mules is concerned, he is much sought after by French breeders and taken to Savoy, where he becomes the sire of most excellent mules for (comparatively speaking) light draught work. By judicious crossing, the writer succeeded, in some four or five generations, in producing jacks with the whole of the desirable qualities above referred to—viz., size, bone, short legs, courage, as well as good general conformation.

The Americans have always attached great importance to height in a jack to the disregard of other qualities; but it is not the tallest jacks that get the best mules. In point of fact, it is very much the contrary, and it is rare to find a jack exceeding 15 hands which can be properly classed as "short legged." Excessive height in a jack necessarily implies height on the leg—a most undesirable point in a breeder's eyes on this side of the water. Jacks of 16 hands high are not uncommon in the States. The tallest jack known is believed to have been a Catalonian imported into Tennessee in 1887. He was called "Great Eastern," and had been awarded first prize at the great show at Puycerda, in Catalonia, in 1886. He stood 16 hands 3 inches high, and was used in Tennessee for jennies only, at a fee of 10*l.* for each, being the same amount as that charged in this country for the services of the Shire horse Prince Harold, recently sold for upwards of 2000*l.* Unfortunately for his owners, Great Eastern became badly "foundered" soon after arriving in the States.

For a number of years teams of large mules have been regularly worked at Badminton, the hunting seat of the Duke of Beaufort, for farm and general carting work. At one time they were used a good deal, in a team of four, in the hound-van, but as they got on in years it was thought that they were not fast enough, and the billet was handed over to old hunters and harness horses. A mixed team of mules in the wheel and old hunters in the lead would have proved successful, and, supposing the old hunters were "quick" enough to get out of the way of the bars, it would have been found that a pair of well-bred mules made most efficient wheelers, "collaring" and "holding" in a way that would quite astonish an orthodox "coachman."

Mr. A. J. Scott, of Rotherfield Park, near Alton, Hants, has also bred a number of large mules (from English cart mares and foreign jacks), which he employs for farming and estate work, and which give great satisfaction. He has also bred several jacks and jennies, which have been exported to various countries for mule-breeding purposes.

The following general facts in connection with the subject under consideration may not be without interest. Mules are commonly sold by weight, unseen, in the United

States. Thus a mule dealer at Philadelphia will telegraph to perhaps the proprietor of the "Mammoth" Mule Yards at St. Louis, the headquarters of the mule trade, to send him a car-load (generally eighteen) of "smooth mules," averaging, say, 1000lb. weight each.

Instances are on record in which mares have given birth to twins, a mule and a horse. These were clearly cases of superfœtation.

The longevity of the mule is one of its chief recommendations. The writer, having made a study of the mule during forty years at home and in various parts of the world, as well as having bred and worked them regularly, is able to affirm that he has never in his experience seen a dead mule, and that he has never gone out of his way to avoid seeing one.

A remark or two on the difference between mules and jennets would not perhaps be out of place. The jennet favours the mare in about the same degree that the mule favours the jack. It is generally supposed that in crossing the donkey is "prepotent" over the horse. In the case of the mule the jack is very prepotent, but it is not so in the case of the jennet, which may be said to be more "half and half." An expert has no difficulty in distinguishing mules from jennets. For trotting work the jennet is the better animal, and he has great power of endurance as well as longevity. The jennet is much bred in Ireland, especially in Limerick, Cork, and parts of Tipperary. In the congested districts, the chief place for breeding jennets is in the neighbourhood of Swinford, in co. Mayo, and in 1894 a very good Welsh pony was stationed there by the Congested Districts Board for the purpose. It is thought, however, that the breeding of jennets is carried on in a very haphazard way in Ireland—with no care, using the worst possible kind of common country ponies as sires, and probably breeding them because the donkeys have become so much deteriorated that they are of very little use. The Irish, at all events, would seem to have benefited by their proceedings. For the above information regarding Irish jennets the writer is indebted to Mr. Frederick Wrench, of the Irish Land Commission. A great many jennets are bred in the neighbourhood of Naples, and also in Sardinia.

It is now proposed to add a few final and practical remarks on the use of mules. It must always be remembered that a mule is not a natural animal, but that he is rather the invention of man. He has been aptly described as an animal with "no ancestry, and no hope of posterity." Brought up by the side of the mare (his dam), he adores the whole horse tribe, and hates the asinine race generally. He is always nervous, and afraid of strangers. Whilst he is a "natural puller," and has enormous strength, he is loth to make use of it to the utmost unless he has a "lead" given him. This "lead" should be always, if possible, a horse, or, better still, a white mare. It seems curious that, while this peculiarity has long been known in Spain and Italy (where the diligence always had a horse of some sort or other in front of the mules to give them a lead), it has only been *partially* recognised in other countries (our own colonies for instance) where the mule has been adopted. Thus at the Cape, in the coaches, a mare is sometimes put to run *by the side* of the mule team. The mare should be *in front* of the whole team, as one of a pair with a mule, but always *in front*. The mules will follow her, and, being creatures of imitation, will do their work much more willingly and with less whip.

The employment of a mare driven in front of a mule

team is the key-note to the satisfactory employment of the mule in general for any kind of draught work. It is, of course, well known that, in camping out, an old white mare, with a bell round her neck, will keep 100 mules from straying. It seems only common sense that, when peculiarities of this kind are so well known, a mare should be always, and not occasionally only, employed for this purpose. It will be found that a pair of mares, or even horses, used as the "first leaders" will produce the same effect, and that pack mules will equally well follow a horse or mare. Exemplification of this principle, or peculiarity, may be seen in the Old World in the streets of Genoa, and in the New World in those of Philadelphia, or almost any great city in the States. At Genoa it takes the more economical form of a donkey in front of a mule, the latter being harnessed to a cart loaded with two or three tons of material. It would be absurd to suppose that the poor donkey is much good, but he does his best in front of the mule, and the mule seeing this puts his best leg foremost, not, however, without some persuasion on the part of the generally brutal carter. At Philadelphia long strings of mules may be seen drawing railroad cars through the city. They are, or were, mostly known as "Lafferty's teams." Each string consists of from twelve to fourteen mules in single file, and each string has a horse or mare in front of the mules, thus recognising the necessity of the mule requiring a "lead." It must not, however, be thought that all mules require a "lead;" they differ very much from each other in this respect. Some will go first and do their work honestly; but it must be accepted as a general rule that, in order to get the maximum amount of work out of a mule team, a horse or mare of some kind should head each team. Neglect of the precaution of always having a small proportion of horses among mule teams may very possibly end in disaster where army transport is concerned.

Some years ago, during the progress of one of the little wars in South Africa, certain "imperial officers" were sent up the country to buy mules for the service. Arrived at a breeding farm, which happened to belong to an educated English gentleman, certain mules were shown which were running in an inclosure with two old ponies, the latter for company's sake. A bargain was struck for the whole of the mules, and it was suggested by the seller that the officers should take the two ponies for an old song, as it might facilitate their getting the mules down to headquarters. The seller was rather curtly informed that their "orders were to buy mules, not ponies." The absence of any practical knowledge of the subject on the part of the headquarters staff is as self-evident as is the want of discretionary power accorded to the purchasing officers. It is thought that the mules are still wandering about the veldt somewhere in South Africa!

The question as to the occasional fertility of mules is an interesting one, and has already been referred to. As a general rule, it may be set down that the mule, both male and female, is absolutely sterile, although the generative instinct is perfectly developed in both sexes. It is not proposed here to enter into a physiological discussion on the subject, but the reader will find the various points *pro* and *con.* admirably discussed in the second part of M. André Sanson's "Economie du Bétail." The so-called fertile mule "Catherine," still existing at the Jardin d'Acclimatation, Paris, may or may not be the exception which proves the rule, but it is necessary, in the first instance, to prove that "Catherine" *is* a mule. From the first it was taken for granted that she was a mule, but her parentage as such has never been properly authenticated. "Catherine" was imported in 1873 from Algeria, and after this lapse of time (twenty-two years) it is found impossible to obtain the necessary information. The fact that her offspring by a horse are fertile, while those by an ass are sterile, tends rather to show that she is merely a mare whose dam once bore a mule, and subsequently bore "Catherine," the latter showing signs of the influence of a previous impregnation. On the other hand, by the casual observer she would at once be pronounced a mule from her general appearance, her style of playing, her walk, her head and ears, and her voice, all of which are mulish. In any case, she is the only instance of a *possibly* fertile mule that has ever come under the writer's observation after a rather wide experience. The various cases which from time to time are reported from the United States must be taken—in the absence of definite information regarding the parturition of the mule, which is never given—as cases of induced lactation. In warm climates it is stated that occasionally female mules become pregnant, but that pregnancy is invariably followed by abortion, and that at an early stage.

The stallion mule is absolutely sterile. He is a most undesirable beast, either in the prairie, park, or paddock. He is, however, much used in Northern Italy for draught work, especially in Genoa. He is capable of performing an enormous amount of work on very little food, but is apt to be a great nuisance in a stable. The sterility of the male mule is allowed on all hands, and if any reader is inclined to question the fact, he is referred to "Annales des Sciences Naturelles" en 1824, tome premier, page 184. In one of the galleries of the museum of the Jardin des Plantes, Paris, there will be found a specimen, two or three days old, of an animal which is labelled as the produce of a mare by a male mule; but the writer was assured by M. Milne-Edwards that too much dependence must not be placed on the statement, as no really authentic information is forthcoming on the subject.

The gelding mule is more generally employed than the stallion, and, as may be readily imagined, is much more manageable and tractable, but does not bring the same price in the market as the female.

The following hint to mule breeders may not be considered out of place. The presence of mules of any age in a paddock or on a prairie where foaling mares are kept should not be tolerated for an instant, supposing that such mares are permitted to foal down in the open. On the birth of the foals, be they mules or horses, they would most certainly be at once killed by the mules out of pure mischief. This often happens in the United States to inexperienced breeders.

In the desultory remarks contained in this chapter, the writer has endeavoured to place fairly before his readers the advantages and disadvantages attaching to the use of mules, derived from practical experience of these animals for many years.

APPENDIX.

MEMORANDUM ON MULE BREEDING.

PREPARED FOR THE USE OF THE GOVERNMENT OF INDIA BY MR. C. L. SUTHERLAND.

General Treatment of Jacks.—In mule-breeding operations it is desirable that, as a rule, the jacks be retained at the *haras* and not sent round the country (although the latter system is undoubtedly more conducive to their health and well being), for the following reasons:

(1) Jacks will often refuse a mare until they have been "prepared" by the presence of a jenny. Another jack, or even a mule, will often produce the desired effect.

(2) Mares will often refuse the jack owing to fear, and require to be teased by a horse and blindfolded.

Some jacks will cover a mare as readily as they will a jenny, and such jacks can be allowed to "travel" as horses do in England; but it will be found that they are the exception.

System in Poitou.—In Poitou, the great mule-breeding district in France, a haras is composed of from four to ten jacks, a stallion horse which covers mares in cases in which it is considered desirable to breed horses and not mules, one or two jennies to excite the unwilling jacks, and one or two horse teasers. One of the latter is ridden daily in the season all round the neighbourhood of the haras to "try" the mares. Those that are found to be in season are, as soon as possible, brought to the haras, where the other teaser is retained for the purpose of further teasing the mares on arrival. A haras with, say, eight jacks will often have a *clientèle* of 600 mares. It is thought that the above reasons are sufficient to warrant the recommendation that the jacks be retained at the haras, and only in special cases allowed to travel round the country.

Exercise or Work.—Each jack should have a separate box, and should have daily exercise, either led or loose, in a *well-secured* paddock. They can be more readily broken to harness and worked in carts than is generally supposed.

Food.—The feeding of all breeding animals requires special attention. All grain which is inordinately rich in fat-forming constituents, as, for instance, Indian corn, should be given sparingly. Taken together, perhaps oats are the best staple food, to which a moderate amount of the leguminous seeds, peas, beans, and vetches, may be added.

Rock Salt.—A lump of rock salt should be placed in each jack's manger; it adds very greatly to the general well-being of the animal.

So-called Vicious Jacks.—There is no jack that is so vicious that he cannot be managed by an expert. Instead of vicious it is better to use the term lively. Some are very lively and frighten people not used to these animals. They will attack and savage a stranger, and take any amount of punishment on the head and body. The Americans have a saying that the "mule is very private and particular about his ears." The same remark applies to the jack. A small twig smartly applied to a jack's ears will keep him off a man better than a thick stick applied to his head or body. No jack will face a birch broom. At the sight of it he will retire to the further corner of his box. To lead a lively jack, get a twitch with a good *thick* piece of rope attached to it. Place the rope in his mouth, *i.e.*, on his lower jaw, and twist it till it is moderately tight. Keep as near the point of the shoulder as possible. If he is extra lively, put on two of these twitches, with a man to each, one on each side of the animal. The length of stick should be from 3ft. to 4ft. This twitch is the severest way of treating a jack, and should be seldom required. A common iron or galvanised iron bit, with cheek pieces from 9in. to 12in. long, will generally suffice to lead and control a lively jack.

Use and Abuse of Sexual Power.—Two leaps per diem from each jack, one in the morning and one in the evening, are all that should be expected, except in very special cases. In Poitou six or seven leaps, up to even twelve, are daily exacted from each animal. The average mule-breeder of France is totally ignorant of the laws of physiology, and has only the love of immediate gain before his eyes. Although this abuse of sexual power does not seem actually to shorten the days of the jack, it materially affects his powers of fecundating his mares. I have known, however, of jacks of twenty-five years of age retaining their fecundating powers in spite of having been grossly abused.

Number of Mares to each Jack.—Looking at the well-ascertained fact that the mare holds less readily to the jack than to the horse, and consequently requires to be served in the generality of cases a greater number of times by the former in order to prove in foal, it is fair to put the number of mares for each jack at from fifty to seventy. In cases in which jacks are intelligently managed and fed this number may be increased to one hundred mares.

Jacks Serving Donkey Mares.—Some jacks will never cover a mare after they have once covered a jenny. The first service should be on a mare if possible, and the jack should not be allowed to serve a jenny until the end of the season, after having served all the mares required. By the beginning of the following season he will have forgotten to a great extent the jennies, and will begin with the mares again. It is only natural that he should prefer his own species. There is a very marked difference in his behaviour and general demeanour when covering mares or jennies. In certain cases it may be desirable to reserve a certain proportion of the jacks for breeding what is called in the United States "jack stock," as it is quite possible to spoil a good mare-server by allowing him to have connection with his own species. These jacks are called "jennet jacks" in the United States, and are specially reserved for the production of jack stock.

In the United States stallion donkeys are called "jacks," mare donkeys "jennets" or "jennies," and the two together are spoken of as "jack stock."

Mode of Exciting a Jack.—The presence of a jenny is the best and simplest, but, failing that, the presence of anything with which the animal has been brought up when young. The means vary with each animal, and it is often a tedious and slow process. Thus a jack brought up with cows, as sometimes happens, will require a horned beast to be present as a *dernier ressort.* A jack I knew in Poitou had been hand-reared by a little girl owing to his dam having been burnt to death the night he was born. This jack always required a *maquignon* or groom to clothe himself with a horse-rug round his legs before he would prepare himself. He was a most excellent mule-getter, but under ordinary circumstances, if transported far away, would have been at once condemned as useless in the absence of the above information. Some jacks are very lethargic, but this failing may generally be got over by allowing them to see another jack perform, when their feeling of jealousy will be aroused, and they will prepare themselves. A jack having been prepared will sometimes require to be lifted on to the mare by two men, each man seizing a fore-leg, and care being taken that he cannot savage the men. He should not be muzzled as a rule. In cases in which the mare is much higher than the jack, the former should be placed in a hole with her head fastened to a strong ring in a post in front, and a quantity of stable dung placed behind and firmly trodden down to the required height, which may in some cases be up to the hocks. This is the usual custom in France, Spain, and Italy. In the United States the jack is raised on a kind of platform, but, having tried both plans, I incline to the former as the better and less dangerous method. It is imperatively necessary that the mare be hobbled. The neglect of this precaution frequently results in broken legs and other injuries.

Rearing Jacks for Mule-breeding.—Looking at the fact that certain Punjabi, Bokhara, Persian, and, in the first instance, Arab donkeys have been considered good enough to use as jacks, it is fair to presume that among these breeds some jennies can be found good enough to continue the race of mule getters when crossed once or twice with the imported European jacks. In Mexico—and, I am informed, in Persia—immediately a jack is born he is taken from the jenny and handed over to a mare to suckle and bring up. This plan requires very considerable care to get the mare to take to the jack foal. It is, however, quite the best. If this arrangement cannot be carried out, the young jack may be reared by his own dam, weaned at six months, and then brought up till he is two or three years old in the constant company of a filly of his own age. The chief thing to bear in mind is that jacks, and in fact all animals, take to whatever they have been brought up with when young.

Conclusion.—In this Memorandum I have endeavoured to point out the practices followed in countries with which I am practically acquainted, and in which the breeding of mules is an all-important rural industry. There may be, and doubtless are, difficulties attending the carrying out of these practices in India, and I must leave the consideration of them to the authorities on the spot.

LIST OF CHIEF PRIZES WON BY FOREIGN MULES AND ASSES

BELONGING TO

MR. C. L. SUTHERLAND,

Down Hall, Farnborough, Kent.

AUGUST, 1864.—Agricultural Hall, London.

Second prize for "La Comtesse d'Abbeville," a French ass imported from Picardy, 13 hands.

JULY, 1865.—Agricultural Hall, London.

First prize for "La Comtesse d'Abbeville."

Second prize for "Malta," half-bred English and Maltese.

Between 1865 and 1873 there were no shows of mules and asses held in England.

JULY, 1873.—Royal Agricultural Society's Show, Hull.

First prize for "Don Pedro II.," a Franco-Spanish stallion ass, 14 hands, bred by exhibitor, as the best jackass for getting mules for agricultural purposes. Dam, "La Comtesse d'Abbeville." (Exported to South America.)

First prize for the best mule for agricultural purposes; "Marshal McMahon," Poitou mule, 16 hands high, five years old.

MAY, 1874.—Crystal Palace Show of Mules and Asses.

Very highly commended and commended for "Rousseau" and "Blossom," Poitou mules, 15.2 and 16 hands respectively, regularly used by exhibitor for farm work.

First prize for "Iago," imported Spanish ass, 14 hands, for improving English donkeys and breeding mules.

Third prize for "Borrico," imported Spanish ass, 13.1. (Both these asses were subsequently exported to Jamaica.)

Second prize for "Prima Donna," Spanish jenny, 13.2. (This jenny has offspring in all four quarters of the globe.)

JULY, 1874.—Alexandra Park Horse Show

First and second prizes for "Rousseau" and "Blossom," Poitou mules, described above.

JULY, 1874.—Royal Agricultural Society's Show, Bedford.

First prize for Spanish ass "Iago," described above.

First and third prizes and reserved number for "Rousseau," "Blossom," and "Robin," Poitou mules, described above.

MAY, 1875.—Crystal Palace Show of Mules and Asses.

First and second prizes, highly-commended, and commended for "Beauty," 17 hands; "Brunette," 16.1 hands; "Baron," 16.1 hands; and "Boinot," 15-3 hands, Poitou mules, regularly used for farm work by exhibitor.

First and third prizes for "Blossom," 16 hands, and "Sweep," 15.1 hands, Poitou mules, regularly used by exhibitor for trotting and dog-cart work.

(Equal) *first prize* for "Lad of Poitou," imported Poitou jack, for breeding heavy draught mules, height 13.1. (Exported to the Cape of Good Hope.)

First and second prizes for "Anesse" (Poitou) and "Prima Donna" (Spanish), the former with pure-bred Poitou jack foal at foot.

JUNE, 1875.—Bath and West of England Society's Show, Croydon

First prize for "Brunette," Poitou mule, 16.1.

JUNE, 1875.—Alexandra Park Horse Show.

First and second prizes for "Beauty," 17 hands, and "Brunette," 16.1 hands, Poitou mules.

JULY, 1875.—Royal Agricultural Society's Show, Taunton.

First prize for "Comte de Poitou," 13.2 hands, imported Poitou ass, for getting mules for agricultural purposes. (Exported to the Cape of Good Hope.)

First and second prizes for "Brunette," 16.1, and "Beauty," 17 hands, Poitou mules, described above.

OCTOBER, 1877.—Dairy Show, Agricultural Hall, London.

First prize, with silver medal, for "Comte de Vitré," imported Poitou stallion ass, 15 hands, for breeding draught mules.

Third prize for "Ranulfe, Comte de Poitou," imported Poitou ass, 13.2. (Exported to Jamaica.)

Second prize for "Prima Donna," Spanish ass, 13.2.

First, with silver medal, and second prizes for "Brunette" and "Beauty," Poitou draught mules, described above, and *commended* for "Bravo," draught mule, 16 hands.

First, with bronze medal, and second prizes for "Centennial Harry," 14.2, light American piebald mule, imported from Kentucky, and "Blossom," Poitou mule, 16 hands.

JULY, 1879.—Royal Agricultural Society's Show, Kilburn.

First, second, and third prizes for "Beauty," "Blossom," and "Brunette," Poitou draught mules, described above.

First prize for "Centennial Harry," light American piebald mule, described above.

First prize for "Comte de Vitré," Poitou stallion ass, above described.

First prize for "Adèle," Poitou jenny ass, 13.2.

JULY, 1881.—Alexandra Park Mule and Donkey Show.

First prize for "Brunette," Poitou heavy draught mule, 16.1.

First prize for "Belle," imported Kentucky trotting mule, 16 hands.

First prize for "Comte de Vitré," Poitou stallion ass, 15 hands.

Highly commended for "The Duke," Spanish stallion ass, imported from Andalusia, 14 hands.

First prize for "Donna II.," Spanish ass, bred by exhibitor, 13.2.

Highly commended for "Adèle," Poitou ass, 13.2.

JUNE, 1889.—Royal Agricultural Society's Show, Windsor.

First prize for "Malta Jack," Maltese stallion ass, 14 hands.

Second prize for "Cetywayo," Poitou-Maltese-Catalan stallion ass, 15.1, bred by exhibitor.

Mr. Sutherland has bred and exported Poitou and Spanish asses, for mule breeding, to Scotland, Ireland, Natal, Cape of Good Hope, South America, Jamaica, Montserrat, the Fiji Islands, the United States, British India, British Honduras, and New South Wales.

INDEX.

African Wild Ass ... ...*page* 11
African Ass Hybrids 14
American Mule 107
Appendix 153
Asiatic Ass Hybrids 67
Ass, Asiatic 21
Ass Hybrids 67
Ass, longevity of 18
Ayrault, M., number of Mules in Poitou 90
Ayrault, M., on infertility in Mules 81

Baggage Mules 131
Baker, Sir Samuel, on African Ass 12
Bardot 100
Bartlett, Mr., production of Bovine Hybrids 79
Baudet 101
Beaufort, Duke of, use of Mules by 145
Blanford, Mr., on Asiatic Ass 22
Browne, General Sir S., on domesticated Onagers 29
Bryden, Mr. H., on the Quagga 62
Bullock Teams 142
Burchell's Zebra 51
Burchell's Zebra, use in Paris 143
Bureau of Animal Industry—Report on Mules 115

Catalan Jacks 116
Catherine, supposed fertile Mule 149
Chapman's Zebra 51
Churchill, Lord Randolph, African Donkeys 142
Churchill, Lord Randolph, misnomer of Burchell's Zebra 60
Clay, Hon. H. B., introduces Catalan Jacks into Maryland 111

Dzeggettai 21

Endurance of Mules in Federal Army ... ...page 125
Equus asinus 11
Equus burchellii 51
Equus caballus, characters of ... 1, 2
Equus, existing species of ... 1
Equus grevyi 43
Equus hemionus ... 21
Equus hemippus 21
Equus przewalskii 7
Equus quagga 61
Equus somalicus ... 19
Equus tæniopus 12
Equus zebra 37
Evans, Mr. J. B., on lactation in Mules 84

Flower, Sir W., on *E. grevyi* 45
Flower, Sir W., organisation of the Horse ... 5
Flower, Sir W., on *E. przewalskii* 8
Food of Breeding Jacks 154
Francis, Francis, lactation in maiden Animals 83
Fossil species of Horse 4

Gestation, period of, in Mare ... 2
Gestation, period of, in Ass 14
Ghorkhur 23
Ghur 23
Gilbey, Sir Walter, notice of Washington's Mules ... 107
Gordon, W. J., erroneous account of Mules ... 73
Grevy's Zebra 43
Grijimailo, the Brothers, on Prejevalsky's Horse 9

Hallen, Col., on Indian Baggage Mules 133
Harrington, Mr. J. L., on riding down Onagers ... 25
Harris, Capt. W. Cornwallis, on mountain Zebra... 39
Harris, Capt. W. Cornwallis, on Quagga 61
Hay, Major W. E., on the Kiang 32
Hayes, Capt., on Burchell's Zebra 56
Hayes, Capt., on markings of Ass ... 61
Hayes, Capt., on non-fertility of Mules... 80

Hayes, Capt., on taming the Zebra ...*page* 40
Hemippe 29
Hinnies 100
Hippotigris 37
Hore, Mr. Fraser S., on Onager 23
Horse, the 1
Horse, distribution of ... 3
Horse Hybrids 66
Horse, period of Gestation 2
Humboldt, Baron, lactation in Man 82
Hybrid African and Asiatic Asses ... 14
Hybrid Equidæ ... 65

Indian Transport Mules 129
Italian Jacks for Mule Breeding ... 136
Irish Jacks 140

Jacks, comparison of various Breeds ... 121
Jacks for Mule Breeding 119
Jennets 119
Jennets, characteristics of 146
"Jennet Jacks" 156
Jones, Mr. J. L., report on Mules 116

Kiang 21, 30
Kentucky Mules 141
Killgore, Mr., on American Mule Breeding 110
Kinloch, Colonel, on the Kiang 30
Kipling, Mr. John L., on character of Mules ... 74
Koulan 23

Lactation in Mule 82
Lactation in sterile Animals 83
Layard, Mr. E. L., on Chapman's Zebra ... 52
"Lead" desirable for Mule Teams 147
Leith, Mr. A. H., on Baggage Mules 131
Leonard, Major A. G., on Mules as Pack Animals ... 127
Lugard, Capt. F. D., on Transport Mules 128
Lugard, Capt. F. D., utilisation of Burchell's Zebra 58

Management of Mule-breeding Mares ... ...*page* 122
Meux's, Sir H., on hybrid Burchell Zebras 52, 66, 143
Morgan, Mr. E. Delmar, on Prejevalsky's Horse 9
Morton, Mr. John Chalmers, on use of Mules 77
Mules, advantages of 75
Mule, American 107
Mule Batteries 129
Mule Breeding in India 153
Mule Breeding in Poitou 153
Mules for Military Service 127
Mules, non-fertility of 78
Mules, lactation in 78
Mules, longevity of 146
Mules, prevalent ignorance regarding 72
Mule, Poitou 85
Mule, Spanish 88
Mules undesirable in Breeding Paddocks ... 151

Neumann on Grevy's Zebra 48
Nutt, Captain H. L., on Onager ... 23

Off-coloured Jacks 138
Off-coloured Jacks, inferior value of 140
Onager 23
Onagers, breaking-in 27
Overworking Jacks 155
Owen, Sir Richard, on relation of Horse to Man ... 4

Pack Mule, required size 141
Punjab Pack Saddle 130
Phillips, Mr. Lort, on Somali Ass 20
Poitou Ass 95
Poitou Ass, measurements of 102, 103
Poitou Asses, breeding of 104
Poitou Mule 85
Poliakof, description of, *E. przewalskii* 7
Practical remarks on Mules 138
Prejevalsky's Horse 7

Pringle, Thomas, on the Quagga ... ...*page* 62
Prizes won by Mr. C. L. Sutherland's Mules and Asses ... 159

Quagga 61
Quagga Hybrids 69

Rayment, Major, on Mule Breeding 137
Rearing Jacks for Mule Breeding 157
Riecke, Mr., on supposed existence of Quagga ... 64
Rock Salt desirable for Mules 154
Rothschild, Hon. Walter, on utilisation of Burchell's Zebra 59

Selous, Mr. F. C., on Burchell's Zebra 57
Sclater, Mr. P. L., on *E. grevyi* 43
Sclater, Mr. P. L., on Somali Ass 19
Scott, Mr. A. J., Mule bred by 93
Scott, Mr. A. J., Mules bred by 145
Smith, Col. Hamilton, on Sexes of Mules 73
Smith, Colonel C. Hamilton, on the Quagga 61
Somali Ass 19
Stephens, Mr. Harold, on Burchell's Zebra for draught ... 54
Sterility of Male Mule 150
Sutherland, Mr. C. L., on distinctions between Ass and Horse 17
Sutherland, Mr. C. L., prizes taken by Mules 89
Supposititious Mule 81
Swayne, Capt. H. G. C., on Grevy's Zebra 45
Syrian Wild Ass 29

Tarpans 3
Tegetmeier, Mr. W. B., on lactation in Mule 83
Thatcher, Mr., on Mule-breeding in China 137
Times, the, on Indian Transport Mules 129

United States, number of Mules foaled in 1889 117

Valentine, Mr. J. Tristram, on *E. grevyi* 47
Value of Jacks in Poitou and Kentucky 139
Vicious Jacks, management of 154

Warder, Mr. J. T., report on utilisation of Mules ...*page* 113
Washington, General, on utility of Mules 107
Weight of Mules 146
Wister, Col. Langhorne, on use of Mules ... 77
Wrench, Mr. Fred., on Irish Jennets 147

Zebra, gestation of 38
Zebra, Burchell's ... 51
Zebra, Grevy's 43
Zebra, Mountain 37
Zeedesberg, Mr. J., team of Burchell's Zebra 55
Zebra Hybrids 67

www.ingramcontent.com/pod-product-compliance
Lightning Source LLC
Chambersburg PA
CBHW021703210326
41599CB00013B/1507

* 9 7 8 3 3 3 7 1 4 2 9 8 8 *